CAMBRIDGE LIBRARY COLLECTION

Books of enduring scholarly value

Physical Sciences

From ancient times, humans have tried to understand the workings of the world around them. The roots of modern physical science go back to the very earliest mechanical devices such as levers and rollers, the mixing of paints and dyes, and the importance of the heavenly bodies in early religious observance and navigation. The physical sciences as we know them today began to emerge as independent academic subjects during the early modern period, in the work of Newton and other 'natural philosophers', and numerous sub-disciplines developed during the centuries that followed. This part of the Cambridge Library Collection is devoted to landmark publications in this area which will be of interest to historians of science concerned with individual scientists, particular discoveries, and advances in scientific method, or with the establishment and development of scientific institutions around the world.

A Sketch of the Physical Structure of Australia

The geologist Joseph Beete Jukes (1811–1869) studied at Cambridge under Adam Sedgwick (1785–1873). In 1841, having already gained field experience in England and Newfoundland, he joined the H.M.S. *Fly* as a naturalist for an upcoming four-year expedition to chart the coasts of Australia and New Guinea. In 1847, he published a two-volume *Narrative of the Surveying Voyage of H.M.S. Fly* (also reissued in this series). That was followed in 1850 by this pioneering study of the geology of Australia, which drew on Jukes' own observations as well as on some earlier work. It describes features including the mountains along the East coast, raised beaches, alluvial plains and the Great Barrier Reef, and rock types from limestone and sandstone to granite and lava. The book made an important contribution to the scientific literature on Australia at a time when that continent was still largely unknown to European scholars.

Cambridge University Press has long been a pioneer in the reissuing of out-of-print titles from its own backlist, producing digital reprints of books that are still sought after by scholars and students but could not be reprinted economically using traditional technology. The Cambridge Library Collection extends this activity to a wider range of books which are still of importance to researchers and professionals, either for the source material they contain, or as landmarks in the history of their academic discipline.

Drawing from the world-renowned collections in the Cambridge University Library, and guided by the advice of experts in each subject area, Cambridge University Press is using state-of-the-art scanning machines in its own Printing House to capture the content of each book selected for inclusion. The files are processed to give a consistently clear, crisp image, and the books finished to the high quality standard for which the Press is recognised around the world. The latest print-on-demand technology ensures that the books will remain available indefinitely, and that orders for single or multiple copies can quickly be supplied.

The Cambridge Library Collection will bring back to life books of enduring scholarly value (including out-of-copyright works originally issued by other publishers) across a wide range of disciplines in the humanities and social sciences and in science and technology.

A Sketch of the Physical Structure of Australia

JOSEPH BEETE JUKES

CAMBRIDGE UNIVERSITY PRESS

Cambridge, New York, Melbourne, Madrid, Cape Town,
Singapore, São Paolo, Delhi, Tokyo, Mexico City

Published in the United States of America by Cambridge University Press, New York

www.cambridge.org
Information on this title: www.cambridge.org/9781108030847

© in this compilation Cambridge University Press 2011

This edition first published 1850
This digitally printed version 2011

ISBN 978-1-108-03084-7 Paperback

A SKETCH

OF THE

PHYSICAL STRUCTURE

OF

AUSTRALIA.

THE EASTERN ARCHIPELAGO.

By Permission of the Lords Commissioners of the Admiralty.

Now ready, in 2 vols. 8vo. with numerous Maps, Plates, and Woodcuts,

NARRATIVE

OF THE

SURVEYING VOYAGE OF H.M.S. FLY,

UNDER THE COMMAND OF

CAPTAIN BLACKWOOD, R. N.

IN TORRES STRAIT, NEW GUINEA, AND OTHER ISLANDS

IN THE EASTERN ARCHIPELAGO;

TOGETHER

WITH AN EXCURSION INTO THE INTERIOR

OF THE

EASTERN PART OF THE ISLAND OF JAVA,

DURING THE YEARS 1842 TO 1846.

BY J. BEETE JUKES, M. A.

NATURALIST TO THE EXPEDITION.

OPINIONS OF THE PRESS.

" We must congratulate Mr. Jukes on the value of his publication. Scientific without being abstruse, and picturesque without being extravagant, he has made his volumes a striking and graceful addition to our knowledge of countries highly interesting in themselves, and assuming hourly importance in the eyes of the people of England."—*Blackwood's Magazine.*

" To transcribe the title-page of this book is sufficient to attract public curiosity towards it—to peruse the book itself is to be rewarded with the knowledge of a mass of information in which complete confidence can be reposed, for, from the first page to the last, it is apparent that the main object with Mr. Jukes is to tell all that he knows and believes to be true, rather than to win favour from his readers by his manner of telling it. There is not a pretty phrase, an exaggeration, nor an invention in the two volumes of Mr. Jukes; all is plain unadorned fact, and because it is so, is deserving, not merely of perusal, but of study. Such are the recommendations of Mr. Jukes' pages to the public, and all who desire to see truth united with novelty will peruse them."—*Morning Herald.*

" Mr. Jukes has been most judicious in his selection of topics whereon to dwell in his narrative, and he describes with great vivacity and picturesque power. There is much novelty and freshness in his book, and much valuable information."

Daily News.

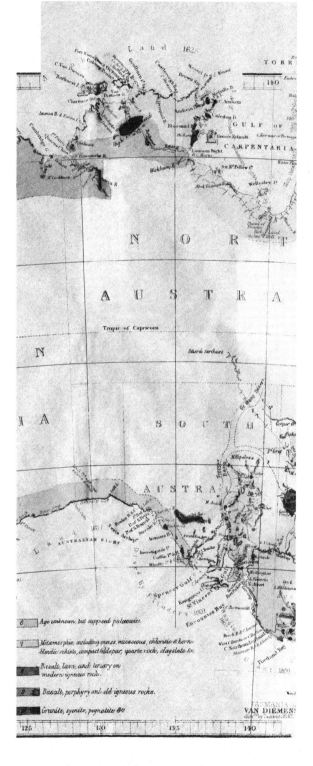

Land 1823

TORR

C. Van Diemen
Bathurst I.
Van Diemen G.
Clarence Str.
Anson B. & Evan I.
Cambridge G.
Queen's Chan.
Port Keats
M.Cockburn
R.

Anhem
Arnhem
Caledon B.
Brue and R.
McTartes
Limmen Bight
C. Maria
Wickham R.
Abel Tasman R.
Wellesley Is.
Groote Eiland

GULF OF
CARPENTARIA

Chani or
Browns
Is. 1841

N O R T

A U S T R A

Tropic of Capricorn

N

Sturt's farthest

I A

S O U T H

Cooper C.

Fr. Howe

P. Grey

M.Hopeless

M.Arden

A U S T R A L

L. AUSTRALIAN BIGHT

Rocky B.

Port Lincoln
Stokes B.
Anxious B.
Investigator Str.
Cape
Whidby I. & C.

Spencer Gulf
St. Vincent
Kangaroo I.
Encounter Bay

Wellington
L.Victoria
L. Albert

Rivoli B. & C.Lacepede
West Banks or C.Banks
C. Northumberland
Discovery B.
Portland Bay

6   Age unknown, but supposed palæozoic.

7   Metamorphic, including gneiss, micaceous, chloritic & horn-
     blendic schists, compact feldspar, quartz rock, clayslate &c.

8   Basalt, lava, and tertiary or
     modern igneous rock.

9   Basalt, porphyry and old igneous rocks.

     Granite, syenite, pegmatite &c.

TASMANIA
VAN DIEMENS
disc. by Tasman 1642.

125   130   135   140

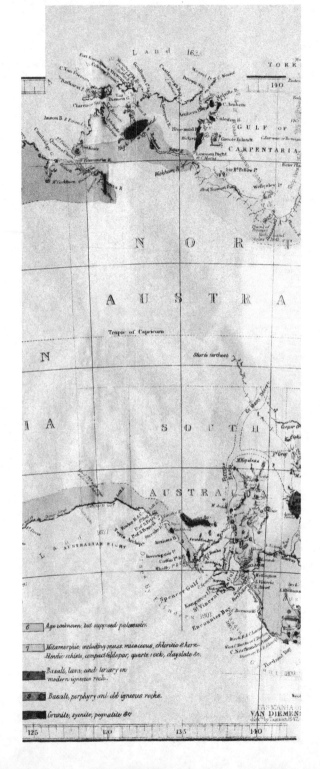

Land 1623   TORR

C. Van Diemen   140
Bathurst I.

Clarence Str   Van
Diemen G.   C. Arnhem

Anson B. & Bron I.   Caledon B.   GULF OF

Cambridge G.   Blue mud B.   CARPENTARIA

M'Cockburn   Wickham R.

Abel Tasman Pt.   Wellesley Is

                    N   O   R   T

        A   U   S   T   R   A

            Tropic of Capricorn

                        Sturts farthest

    N

A   S   O   U   T   H

                    A   U   S   T   R   A

                                AUSTRALIAN BIGHT

            L a n d   1671

6   Age unknown, but supposed palæozoic.

7   Metamorphic, including gneiss, micaceous, chloritic & horn-
    blendic schists, compact feldspar, quartz rock, clayslate &c.

    Basalt, lava, and tertiary or
    modern igneous rocks.

9   Basalt, porphyry and old igneous rocks.

    Granite, syenite, pegmatite &c

                                    TASMANIA or
                                    VAN DIEMENS

125   130   135   140

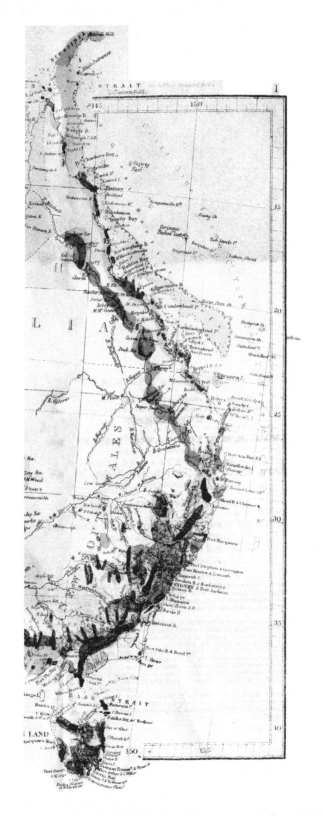

A SKETCH

OF THE

PHYSICAL STRUCTURE

OF

AUSTRALIA,

SO FAR AS IT IS AT PRESENT KNOWN.

BY

J. BEETE JUKES, M.A. F.G.S.

LATE NATURALIST OF H. M. S. FLY.

―――――――

LONDON:

T. & W. BOONE, 29, NEW BOND STREET.

1850.

TO THE

REV. ADAM SEDGWICK, M.A. F.R.S.

WOODWARDIAN PROFESSOR IN THE UNIVERSITY OF CAMBRIDGE,

&c. &c. &c.

MY DEAR SIR,

IT is but a trifle I am going to offer you; but, such as it is, permit me to dedicate this little book to you. It will, at all events, enable me publicly to acknowledge your great kindness to me,—kindness first spontaneously shewn to an idle undergraduate at Cambridge, and subsequently renewed at many periods of my life. Upon the course of that life your influence has been as great as beneficial, and while it lasts you will always be most gratefully and affectionately remembered by

Your sincerely attached pupil,

J. BEETE JUKES.

MUSEUM OF PRACTICAL GEOLOGY,
LONDON, MAY, 1850.

PREFACE.

It is now four years since the notes were written on which this little book is founded. I read an abstract of it at the meeting of the British Association, at Southampton, in 1846; another abstract was read at the Geological Society, in November, 1847. In consequence, however, of its length, and of its being not entirely composed of original observations, it could not be published in extenso in the proceedings or transactions of either of those bodies. I have since frequently intended to publish it myself, in order that the labour I expended on its compilation might not be entirely thrown away; but have been too much occupied with the duties of the Geological Survey of Great Britain, to which I am now attached, to spare the necessary time for its completion. Buried for the most part in the heart of the Welsh mountains, I have been unable to do more now than take advantage of some rainy days this spring for revising my original rough notes. I mention these facts in some measure to excuse the incompleteness of many parts, and the hastily-written style of all, and trust the reader will pardon me accordingly should his eye or his ear be offended.

INTRODUCTION.

I FEAR it will appear to most persons that an attempt to give any tolerable account of the geological structure of so large a country as Australia, from the few observations that have been as yet made and published concerning it, must be a very rash one. This must be especially the case with geologists, accustomed only to the full, varied, and complex structure of western Europe. It is quite clear that any such account must be very incomplete and imperfect. Still I think that sufficient materials have been collected to give a connected outline, a sketch of the structure of that country, which shall exhibit its principal features in a rude and approximate way. My own reasons for attempting such a sketch are, that having visited some portion or portions of all the coasts of Australia, and having made geological notes upon them, I was enabled to understand and connect the observations of others, and that by studying and comparing all the observations I could find in any published accounts, I conceived a general but distinct notion of

B

its structure in my own mind. This notion I shall endeavour to convey to the reader. If it should eventually turn out that my speculations and generalizations are incorrect, no one will be more rejoiced than myself at the acquisition of the more accurate and fuller knowledge that shall overthrow them.

In the meantime I would remind the reader that detached observations have a far greater relative value in such a country as Australia than they would have in a complicated country like Europe. When mountain chains and sheets of rock stretch for hundreds of miles without an interruption, or a change in their general character, cursory observations at great distances one from the other suffice to give us an outline of the country. In Russia, in N. and S. America, as in Australia, observations made across one line of country would thus be applicable to great tracts, while in western Europe a series of observations along any one line might give no information as to the rocks or formations to be met with, even a few miles on each side of it. Australia especially seems the very land of uniformity and monotony, the same dull and sombre vegetation, the same marsupial type of animals, spread over the whole land from the gloomy capes of the south coast of Tasmania, and the stormy Leeuwin, to the cloudless and burning skies of Torres Straits and Port Essington. The reader will, I think, see reason in the following pages to believe that the rock formations of Australia are equally uniform and extensive.

In the colouring of the two maps that accompany this volume, I must warn the reader not to expect any accuracy of outline. They are not geological maps in the strict sense of the word, the geological colours being only dabbed on roughly about the place where the rock indicated by it was observed, as nearly as I could guess its size and locality. The colours are only intended to come in aid of the description, and to *guide the eye* and *help the memory* of the reader.

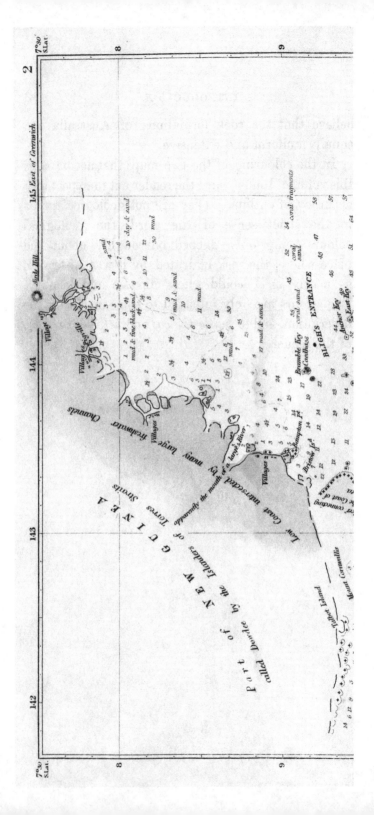

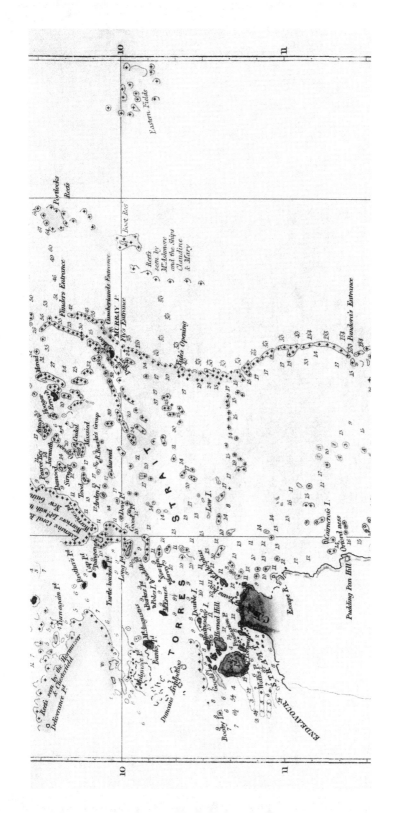

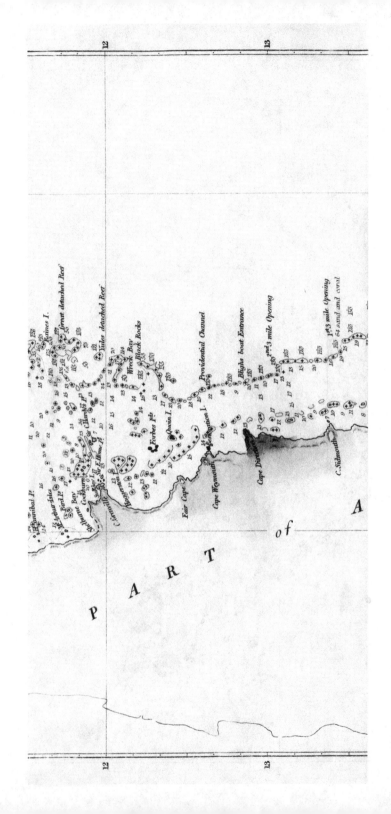

Great detached Reef

Raines I.

Yules detached Reef

Wreck Bay

Black Rocks

Forbes Ps

Owen I.

Providential Channel

Bligh's boat Entrance

1st 3 mile Opening

2nd 3 mile Opening

64 sand and coral

C. Grenville

Montgomery

Sir E. Home I.

Sir C. Clark N.E. P.

Hannibal P.

M. Arthur Isles

Bird I.

Home Bay

Cockburn I.

Fair Cape

Cape Weymouth

Cape Direction

C. Sidmouth

P   A   R   T   of   A

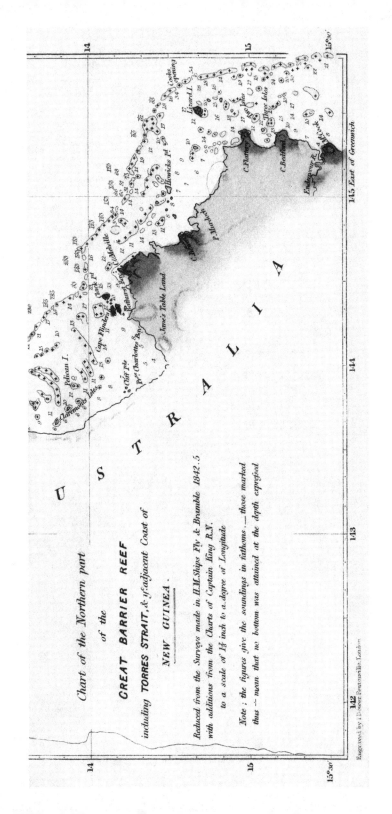

Chart of the Northern part

of the

GREAT BARRIER REEF

including TORRES STRAIT, & of adjacent Coast of

NEW GUINEA.

Reduced from the Surveys made in H.M.Ships Fly & Bramble 1842.5
with additions from the Charts of Captain King R.N.
to a scale of 1¼ inch to a degree of Longitude

Note; the figures give the soundings in fathoms.— those marked
thus — mean that no bottom was attained at the depth expressed

AUSTRALIA

145 East of Greenwich

PRINCIPAL PHYSICAL FEATURES.

THE great physical features of Australia, as far as they are at present known, may be very briefly described.

1. A long, but not lofty, mountain chain runs along the whole eastern coast, crossing Bass's Straits into Tasmania, and running under Torres Straits to the shores of New Guinea.

2. On the landward, or western side of this chain are great plains, declining gradually to the west, but at first often broken by detached hills or groups of mountains. These plains are traversed by the rivers Murray and Darling and their tributaries, on the south, disemboguing into Lake Alexandrina,— by Victoria River (and perhaps some others) in the centre, which drain into Sturt's central desert, and in seasons of flood probably disembogue by Lake Torrens,—and by the small rivers which, on the north, run into the south-eastern portion of the Gulf of Carpentaria.

3. West of the tracts thus described appear immense desert plains, which seem to extend to the sea coast round the Gulf of Carpentaria on the

north,—to that of the Great Australian Bight on the south, and to stretch along the N.W. coast, from Northwest Cape to Collier Bay.

As minor, but sufficiently important features, may be mentioned,—

4. The mountain chain of South Australia running north from Cape Jervis to the singular horse-shoe shaped depression of Lake Torrens.

5. The high land of Western Australia, running north from Point D'Entrécasteaux and King George's Sound to the neighbourhood of Shark Bay.

6 The high land which forms the coast from Collier's Bay to Wickham's Victoria River, and seems to stretch in an east and west direction across the interior of Arnhem Land, south of Port Essington to the western shores of the Gulf of Carpentaria.

I shall endeavour to describe the structure of each of the districts above named somewhat in the above order.

I.—THE EASTERN COAST.

The mountain chain of the eastern coast is, apparently, a rather irregularly formed and complex one, but, it seems to preserve the same or very similar features throughout its extent. It is no where, so far as is known, a single ridge of mountain, but is made up of many masses of various

character,—sometimes peaked and serrated ridges,
—sometimes detached hills rising from slightly
elevated ground,—sometimes great table lands,
often ending towards the sea in nearly perpendicular
escarpments,—sometimes having on one side or
other gently sloping plains furrowed by innumerable
precipitous gulleys and ravines. Large lateral
spurs often diverge on either hand from the central
portion of the chain, and, especially on its seaward
side, appear like separate and independent mountain
masses.

Tasmania may be said to be entirely occupied by
this chain, as it is a complete net-work of ridges
(called there " tiers") enclosing a multitude of small
plains and valleys. In at least two places the
summits of these ridges attain a height of upwards
of 5,000 feet,* while several reach that of 4,000.
From the N.E. corner of Tasmania the chain
may be traced across Bass's Straits by the curved
line of lofty and rugged islands that lead up to
Wilson's Promontory. From this point the chain
strikes into the interior of New South Wales, and
shortly attains its greatest elevation in the " Australian Alps." The height of the loftiest peak of this
group, named by Count Strzelecki† Mount Kosci-
usko, is stated by that observer to be 6,500 feet.

* See Strzelecki's Physical Description of New South Wales
and Van Diemen's Land, pp. 41, *et seq.* He estimates the mean
height of the water-shed of Tasmania at 3,750 feet.

† See as above.

These are the only mountains in all Australia that are
covered by perpetual snow ; from which it follows
(as a necessary consequence in that climate*) that
the only perpetually flowing river of the country,
the River Murray, rises at their feet. The chain
runs thence through the whole of New South Wales
nearly parallel to the coast, or about N.N.E., its
water-shed preserving a mean distance from the
sea of about eighty or one hundred miles. It
sends out, however, many lateral spurs, such as
those which on the land side divide the head-waters
of the rivers Murray, Murrumbidgee, Lachlan and
Macquarrie, the Namoy, the Gwydir, and the
Karaula, and on the sea-side separate the respective
basins of the districts of Gipps' Land and Twofold
Bay, the Shoalhaven River, the Hawksbury and
Hunter Rivers, the Port Macquarrie basin, that of
the Clarence River, and that of the district of
Moreton Bay. Throughout this part of its course
the elevation of the highest points of the main chain
varies from about 2,000 to about 4,000 feet above the
sea.† About Sandy Cape, or in S. lat. 25° the coast

* We might, indeed, say, in that or any other climate, as with
the exception of the St. Lawrence and the Mississippi, and possibly
the Niger and Senegal, all the other great rivers of the world
spring from mountains covered by perpetual snow.

† Count Strzelecki (see above) gives 3,500 feet as the mean
level of the water-shed in New South Wales, which I should have
thought too great by 1,000 feet, had I not known his great
accuracy in observation, and the multitude of observations which
he made.

line changes its direction from N.N.E. to N.N.W., which latter bearing it maintains to its termination at Cape York in S. lat. 10° 43'. Through this northern portion of its course the constitution of the chain is only known from observations made upon the coast, and those of Dr. Leichhardt in his over- land journey from Moreton Bay to Port Essington. Numerous ranges of considerable magnitude, having a north and south course, strike out upon the coast with many bays and indentations behind their head- lands. There are, also, lofty islands stretching off the coast to a distance sometimes of thirty or forty miles. In the interior, Leichhardt makes mention of many mountain masses, whose waters between the parallels of 25° and 18°, all seem to flow down to the eastern coast. Mount Pluto, described by Sir T. Mitchell in his last expedition, in lat. 25°, and long. 147° 30', or nearly 200 miles from the coast, seems to be the most distant source of some of these waters, as the M'Kenzie River of Leichhardt, and the River Beylando of Mitchell, or Cape of Leichhardt; and from it also proceeds Mitchell's Victoria River, running into the interior. North of these, however, the water-shed seems again to approach the coast line, as the head waters of the Burdekin and Separation creek, about 18° and 19°, are not at a greater distance from the sea than 100 miles, and the river Lynd which Leichhardt struck on near lat. 18° flows to the N.W. into the Gulf of Carpentaria. In this neighbourhood also, or between Cape Upstart and Cape Melville, the loftiest and most massive mountain ranges strike out upon the very coast as if the heart or central axis of the chain was here actually washed by the sea. Thus far several of the highest points still attain the height of 4,000 feet above the level of the sea, and its mean elevation cannot be much less than in New South Wales. North of Cape Melville the chain rapidly diminishes in height and importance, few parts being higher than 1,000 feet, and at Cape York we have only insignificant hills of 300 or 400 feet high. As in Bass's Straits, however, so also in Torres' Straits, the submarine continuation of the chain may still be traced in a line of steep, rugged, and often peaked islands, several of which are 7 and 800 feet high, which run in a north and south direction up to Mount Cornwallis, which is close to the coast of New Guinea. The distance in a straight line between the S. Cape of Tasmania in 43° 35′, and Mount Cornwallis in 9° 45′, is 43° 35′ of latitude, or about 2050 geographical, 2370 statute miles; following the slightly curved line of the chain itself, its extreme length will be about 2,400 geographical, or 2,780 statute miles.

We have now to describe the geological structure of this chain. It cannot be said to have any one well-defined axis of granitic or other rocks, or if it has, that axis is often deep-seated, and is not apparent at the surface. Large granitic masses, however, shew themselves at different portions of its course, sometimes forming whole mountain groups, and there is probably no very considerable portion of the chain without the exhibition of granitic rocks, either in its central ridges or on some of its flanking or lateral ranges. Gneiss and mica slate do not seem to be any where abundantly exhibited, although they not unfrequently occur, together with considerable masses of clay—slate and other rocks more or less metamorphic. Great masses of porphyry and greenstone are often associated with these, especially in the lateral ranges.

Upon these rocks lies a large palæozoic formation, which is principally composed of very thick masses of sandstone interstratified occasionally with beds or groups of shale and some beds of limestone. In this palæozoic formation, perhaps only in a particular part of it, beds of good coal are found. Masses of porphyry, greenstone, basalt, and other igneous rocks, are associated with the palæozoic formation, often cutting through, dislocating, and altering various parts of it, but sometimes seemingly supporting it as if existing before its production.

Tertiary rocks again are found reposing at various places on all the above-mentioned rocks, up to a certain height above the sea, and these are likewise associated with still more recent igneous rocks, which sometimes retain all the characters of veritable subaerial lavas.

A. *Tasmania.*—Granitic and metamorphic rocks are described by Count Strzelecki as existing in large mass in the N.E. corner of Tasmania, and in still larger force over its S. W. portion. They shew themselves also at three other smaller districts in the centre and on the northern coast. The rest of the island seems to be generally formed of great intersecting ridges of greenstone, enclosing plains or valleys composed of the palæozoic formation. Some of the more massive greenstones are evidently anterior to the palæozoic formation, as sections may be seen in which horizontal beds of sandstone abut against cliffs of greenstone, or even repose upon steps and ledges of that rock. In other cases, however, sandstones of the palæozoic formation appear to be capped by immense masses of greenstone, and they are certainly often broken through by basaltic and other igneous rocks. The following details on the geology of Tasmania are extracted from a paper of mine in Journal of Geological Society for 1847.

The Valley of the Derwent River.

Along the S.W. side of the valley of the Derwent runs a bold range of flat-topped hills, of which one of the principal promontories is Mount Wellington, rising immediately behind Hobarton to a height of 4,200 feet above the sea. The upper portion of this range is composed of massive greenstone, often forming rude columns of great size, frequently as much as ten feet in diameter. The lower slope of this range, and much of the country forming the opposite side of the valley, is composed of the palæozoic rocks. These lie generally in a nearly horizontal position, and I believe *abut horizontally against the greenstones;* but as I never found a clear section near the junction of the two, I cannot positively say that they do not pass under them,—that the greenstones of the hill-tops are not a thick capping resting on the palæozoic forma tion. In ascending Mount Wellington from Hobarton we first pass over a great thickness of white and yellow sandstones nearly horizontal ; above these are shales and thin beds of limestone, likewise horizontal; over which again other sandstones are found. These rocks occur to a height of 2,500 feet above the sea, and apparently form a solid mass of that thickness at least. Above this point greenstone alone is to be seen, forming a mass 1,700 feet thick at least. Its total thickness depends of course on the undecided question, of whether it be a capping to the palæozoic rocks, or what I believe is more probable, a solid mass with the sedimentary beds resting against its sides.

Both the sandstones and limestones are quarried at several points. At Mr. Hull's limestone quarries at Tolosa, about four miles from Hobarton, I found dark grey limestone, sometimes compact, sometimes finely laminated, with fragments of shells and corals. The beds of limestone were about two feet thick, and in one place were some beds of soft brown sandstone interstratified with thin beds of limestone. These sandstones were scarcely consolidated, and fell to pieces on being taken from the quarry. They often contained fossil shells, both Spiriferæ and Productæ, quite perfect in appearance, but so much decomposed as not to bear extraction, falling into white powdery fibrous carbonate of lime. I procured from other parts of these quarries the following fossils :—

Fossils from Mr. Hull's Quarries.

Corals.

Stenopora Tasmaniensis.	Fenestella internata.
—— informis.	—— fossula ?
Fenestella ampla.	Caryophyllæa.

Mollusks.

Producta rugata.	Spirifer Stokesii.
—— brachythærus.	—— Vespertilio.
Spirifer subradiatus.	—— avicula.
—— Darwinii.	Pecten squamuliferus.
—— Tasmaniensis.	—— Limæformis.

A few miles above New Norfolk, the banks of the Derwent shewed cliffs consisting of alternations of sandstone with black and brown shales, producing a precise resemblance to parts of the English coal-measures. Much fossil wood, apparently parts of large trees, lay in these rocks.

Similar rocks to these were frequently observed in the cuttings of the road-side as far as Oatlands in the centre of the island, and they almost invariably lay in positions so nearly approaching horizontality, that their dip was not appreciable to the eye. Still their continuity did not appear to extend unbroken over any large district, as not only were dykes and other masses of intrusive trap rocks frequent, but solid ridges of crystalline greenstone often intervened, and evidently cut off one portion of the palæozoic rocks from the other.

In the immediate vicinity of Hobarton there were places, as near Stoke, and at the mouth of the valley of Risdon, where the palæozoic rocks had evidently been tilted up and altered by masses of trap rock, which could be traced to have a perfect passage from compact tabular or amorphous basalt into hills of solid crystalline greenstone.

In other places quarries were opened in sandstones of the palæozoic age, forming small patches either embosomed in greenstone, or resting upon it. About a mile from a place called Ralph's Bay Neck, on the S.E. side of North Bay, I found a cliff where the sandstones were shewn clearly to be posterior to the igneous rock. In this case a dark, rudely columnar trap rock ended in a succession of small cliffs and terraces in one direction, upon which terraces and against which little cliffs rested the sandstone perfectly undisturbed, and evidently in the position in which it had been originally deposited.

A parallel instance was observed in the cliffs a little to the eastward of the entrance of Port Arthur.

It appears, then, that there are masses of greenstone both of more ancient and more modern date than the palæozoic rocks

At Macquarrie Plains, about ten miles above New Norfolk, there is a large exhibition of igneous rock, which from its cellular character seems certainly to have flowed as lava in the open air. It forms a mass of considerable thickness, as shewn in the brooks and ravines, and appears to have been gradually accumulated by successive accessions of melted matter. I infer this from the fact of its including fossil trees, apparently in the position of growth, which seem to have been enveloped, while living, in the lava.

There are two small patches of tertiary travertinous limestones : one mentioned by Mr. Darwin, and found in the outskirts of Hobarton, where it appears to have been tilted by the intrusion of an adjacent mass of trap ; another in a little cove called James's Bay, about three miles above Hobarton, on the opposite side of the Derwent. It rests here nearly horizontally, and is but little elevated above the level of the sea. A Helix and a Bulimus, and the leaves and portions of the stems of several plants, have been found in each locality.

Fossils from James's Bay.

Plants, unnamed : one figured by Morris.

Helix.

Bulimus.

There are very thick masses of gravel, consisting of pebbles as large as the fist, accumulated on the sides of the Derwent River at some places, and Count Strzelecki mentions great accumulations of loose sand, from beneath which he procured a large Cypræa. This was at Newton, a short distance from Hobarton.

B. *Norfolk Bay and Tasman's Peninsula.*

The principal mass of Tasman's Peninsula appears to be columnar greenstone, forming the highest and most rugged of its hills, and the gigantic perpendicular cliffs of Cape Raoul and the intermediate shores round the entrance to Port Arthur. Just to the eastward of the mouth of that harbour, a mass of the sandstone of the palæozoic formation, a quarter of a mile across and 200 feet high, may be seen resting against these perpendicular cliffs of columnar greenstone with its beds quite horizontal and apparently unaltered.

Point Puer, one of the projections inside the port is composed of a white compact, rather argillaceous sandstone, which among others contains the following fossils :—

Fossils from Point Puer.

Producta rugata.	Pterinea macroptera.
Spirifer subradiatus.	Orthonota compressa.
—— crassicostatus, *MSS.*, sp. n.	Allorisma, *n.s.*
—— Stokesii.	Pachydomus carinatus.
—— Vespertilio	Pecten squamuliferus.

Eagle Hawk Neck, the connecting link of Tasman's and Forrester's Peninsulas, is one of the celebrities of Tasmania, on account of the peculiar jointed structure of its rocks, forming what is called "the tessellated pavement." The rock is a very hard, brittle, fine-grained and compact grey sandstone or gritstone, lying in a horizontal position. It occasionally contains pebbles of granite, porphyry, or quartz rock.

The rocks abound in fossils, especially at the south point of Pirates' Bay. Among others I collected fine specimens of the following :—

Fossils from Eagle Hawk Neck.

Fenestella internata.	Spirifer subradiatus.
Producta rugata.	—— Vespertilio.
Spirifer crebristriatus.	Platychisma Oculus?
—— Darwinii.	Pachydomus carinatus.
—— avicula.	

On the opposite side of Norfolk Bay is a small peninsula about three miles across, in which is a large convict-station called The Mines. The mass of this piece of land consists of sandstone with some trap, but immediately at the back of the station is a small colliery. A bed of coal, of slight thickness and extent, is here worked. The following was the shaft-section as given me by the overseer :—

	Yards.
"Ironstone" (a fine-grained trap rock) .	20
Sandstone	20
Sandstone and shale	10
Coal	1½

This coal, which in the deepest part is about seven or eight feet thick, rises pretty rapidly in every direction from that point, and as it rises it thins out to about two feet. It thus forms a small basin, not half a mile across, and its outcrop is everywhere covered by beds of loose sand. A little beyond its outcrop on the sea-shore was the following section :—

	Yards.
Trap (in small prismatic pieces) . . .	7
Sandstone, formed of grains of some trap rock .	18
Sandstone, soft and rather shaly . . .	6
Shale and bind	2
Coal	0½

Near this spot they had bored to a farther depth of nearly 100 yards and passed through one twenty-inch coal ; but the rest of the mass was almost entirely sandstone. I got from these coal-measures fossil plants, among which were *Pecopteris Australis,* a Sphenopteris and a Zeugophyllites.

There are other places in Tasmania where coal is worked, but they are chiefly detached and isolated spots separated by green-stone ridges one from the other. I was not able to visit any other of these localities, but I should fear that the beds of coal in Tas-mania are comparatively insignificant in an economic point of view, that the true coal-measures of the country have no great thickness, and that the seams of coal contained in them are but partial, thickening and thinning out perhaps along the same hori-zontal lines, and thus forming limited cakes rather than regular and persistent beds.

C. *East Coast of Tasmania.*

Rocks of the palæozoic formation, chiefly sandstones, are found at various point of the eastern coast, but greatly broken and ob-scured by the usual greenstone ranges and local exhibitions of other trap rocks. In Maria Island are limestone quarries which I did not visit, but from which I procured fossils, among which were some of the large Pachydomi, of precisely the same species as those from Wollongong in New South Wales.

At Spring Vale, about ten miles above Great Swan Port, is a

patch of palæozoic rocks, not more than a mile or two in extent, forming a low gently undulating ground surrounded by hills of igneous rock. No section is exhibited, but blocks of the rock protrude through the soil. It is a fine compact quartz rock, charged with the usual fossils of the formation in great abundance. The rock reminded me strongly of the quartz rock of the Lickey Hill. The fossils of this locality were—

Fossils from Spring Vale.

Fenestella ampla.	Spirifer Stokesii.
Producta rugata?	—— —— crassicostatus, sp. nov.
Spirifer radiatus.	—— three others.
— —— Darwinii.	Stem of a Crinoidal animal.
—— Tasmaniensis.	

B.—The islands of Bass's Straits appear to be chiefly granitic.

C.—*New South Wales.*

Wilson's promontory is all granite, and granite and metamorphic rocks appear from Count Strzelecki's description to extend thence to the Australian Alps, which are almost or entirely composed of them. To the east of this part of the chain occurs the low land of Gipps's Land, which appears to be chiefly tertiary. On the west of the chain palæozoic rocks, with coal, occur around Western Port, and are found in highly inclined positions at some points around Port Phillip where horizontal tertiary strata rest upon them. About Goulbourn and Bredalbane Plains the palæozoic rocks appear to stretch across the whole chain, but farther north granite and metamorphic rocks again appear at the surface, forming a wide district round the head waters of the river Macquarrie, and run thence in a narrower band

to Liverpool plains, whence they again expand to
the north, and occupy a wide band of country as
far north as lat. 28°.* On the western or inland
flank of the chain, limestone and other rocks of the
palæozoic formation are mentioned by Sir T. Mitchell as occurring about Yass plains near the
sources of the Murrumbidgee, between the heads of
the Lachlan and Macquarrie rivers, and about the
head of the Namoy or Peel river. On the eastern
flank of the chain, the rocks on the coast of Twofold Bay appeared to me to be palæozoic, in which
case they probably form part of a large tract of
such rocks spreading over that district. North of
this a granitic spur from the main chain with
porphyry and other igneous rocks reaches the coast
about the Shoalhaven river. North of that again
is the largest and best known district of palæozoic
rocks in New South Wales, stretching from Illawarra to the flanks of the Liverpool range, and traversed by the rivers Nepean and Hawksbury on
the south, and the Hunter river on the north, with
their various tributaries.

From the descriptions of my friend the Rev. W. B. Clarke,†
of Sydney, and from some excursions made either alone or in his
company, as well as from the published accounts of Count
Strzelecki, Mr. Darwin, and others, I am enabled to give a more

---

* Strzelecki's Physical Descriptions, &c.
† I have recently been pleased to hear from Mr. Clarke, that
the Legislative Council have voted him a sum of money to publish
his observations on the Geology of New South Wales in the
colony.

definite account of this district of New South Wales than of any of the others. Immediately north of Illawarra, and on a line running thence nearly due north, about forty miles, to Campbeltown, we get the following section.

In the descending order.

1. Black and brown shales (named, I think, by Mr. Clarke, Wyanamatta shales,) 300 ft. and upwards.

2. The Sydney sandstone (or Hawksbury sandstone), thick sandstones, with a few thin beds of shale in its upper and lower parts; about 700 or 800 ft.

3. Alternations of sandstone and shale ; about 400 ft.

4. Alternations of sandstone and shale, with much fossil wood (often drifted) and some beds of coal ; 200 or 300 feet.

5. Wollongong sandstones, with calcareous concretions, containing many fossil shells and corals, and some fossil wood ; 300 or 400 feet.

Total, 1,900 to 2,200 ft.

The beds here enumerated are probably but a portion of the palæozoic formation of Australia, as there may be found in other places beds much above No. 1, and there are almost certainly beds lower than No. 5.

In these four upper divisions, No. 1 to 4, the fossils are chiefly vegetable, containing among others :

Glossopteris Browniana.	Vertebraria indica.
Pecopteris australis.	Phyllotheca australis.

There are also, I believe, in No. 1, fossil fish to be found, which have not yet been examined so as to determine their genus and species.

In No. 5, the Wollongong sandstone, which is frequently calcareous, and has large and small calcareous nodular concretions, I collected the following fossils :

Stenopora crinita.	Pachydomus carinatus.
Producta rugata.	———— ovalis.
Spirifer subradiatus.	Orthonota, sp. nov.
——— Stokesii.	Pleurotomana Strzeleckiana.
——— avicula.	Bellerophon contractus, sp. nov.

From other localities, whose precise geological or geographical place I do not know, the Rev. W. B. Clarke had, in addition to the above, collected the following fossils :

Favosites gothlandica.	Two species of leptæna.
One species of Crinoidea, appa-	A terebratula.
rently related to Platycrinus.	An eurydesma.
A form belonging to the radiata,	An inoceramus.
and resembling an Echinoderma.	A pleurotomaria.
A small trilobite.	A conularia.
Two new sp. of Spirifer.	

Some persons have been struck with the oolitic aspect of the fossil plants collected in New South Wales (as also of those of India), and have been led to imagine, in consequence, that they did not belong to the same formation as that in which the productæ, spiriferæ, &c. above named, are found. All the physical characters and relations of the rocks, however, both in New South Wales and Tasmania, led me to look upon the whole series as one great continuous formation, and Mr. Clarke has since distinctly informed me, that he has obtained the same spiriferæ, productæ, &c. from beds above those which contain the fossil plants as are found in the beds below.

The perfect conformability and apparent passage from one group into the other would of itself render highly improbable any such difference of age between the higher and lower beds as exists between any palæozoic and any oolitic formation. As to the actual age of these beds, or their identification with any English or European formation, I should not be inclined to express any hasty opinion. Every one who sees the fossils, the long-winged Spiriferæ, the Productæ, and others, must of course be struck with their resemblance to those found in the Devonian formation of Europe. Even were the species identical, however, in rocks thus found at the two opposite sides of the globe, the strict synchronism of the two formations would not therefore be absolutely proved. Many questions, such as the original centre of production of the animals, and the time necessary for their migration and general spreading over the globe,—whether they might not have become extinct at the spot where they first began to exist before they could spread in vast numbers over the opposite hemisphere,— would have first to be settled. Again, if the species are not identical, but only representative, is the fact of that representation always to be taken as proof of strict synchronism ? May not species representing our Devonian have inhabited the seas of the opposite hemisphere, while our own seas were filled with the Silurian animals on the one hand, or the Carboniferous on the other?* Questions like these have yet to be answered before we can determine whether or no strict synchronism can be deduced from the fact of the fossils in opposite hemispheres being representative of, or even if it ever turn out so, identical with each other. At all events, we must not trust too implicitly to single or isolated facts. We must get the series of formations in each case, and compare them with each other, see whether the changes in the one answer to those in the other, and endeavour to trace out some common starting-point of time, before we shall be able to draw clear geological horizons, establish definite chronological epochs common to the whole earth.

Keeping these cautions in view, I should for the present hold the rocks of Australia now under consideration simply as *palæozoic*, and only assert that their age was included within that of our Silurian, Devonian, and Carboniferous periods.

In the district now described, and which may be called the Sydney district, including the counties of Cumberland and Northumberland, and parts of the adjacent counties, the palæozoic rocks have a rudely basin-formed arrangement. About Illawarra they dip to the north, the coal-measures, exposed in the escarpment overlooking Wollongong, coming down to the level of the sea at Bulli, plunging underneath it, and not again emerging till we reach the Hunter River, where they may be seen dipping to the S. The Sydney sandstone above them (No. 2) forms all the coast line between those points, being generally horizontal, but

* There is a certain resemblance between the fauna and flora now living in Australia and those found fossil in our oolitic rocks.

near Sydney having a slight dip to the west, which brings in
(No. 1) the upper shales in the low undulating country occupying
the centre of the county of Cumberland. From that district the
sandstone rises again towards the west, on to the flank of the Blue
Mountains, in the gulleys of which the coal-measures below are
again exposed, while towards the north, I believe, it forms a wide-
spread surface of rock and valley, its beds being in a horizontal
position till we approach the Hunter River, when they rise to the
north, and allow the coal, &c. to appear from beneath them.

Through the whole district under consideration, this Sydney or
Hawksbury sandstone is the most striking and conspicuous rock.
It is usually very scantily covered with soil, and is almost invariably furrowed by innumerable long narrow winding gulleys and
ravines with perpendicular sides, over which the more massive and
harder beds often project in the form of overhanging ledges.
Wherever this rock acquires any elevated position, as on the flank
of the Blue Mountains, this worn and deeply trenched character
becomes remarkably pronounced. I once had a good view of the
gently sloping plateau of sandstone forming the slope of the Blue
Mountains due west of Sydney, from the top of a lofty rock (called
Gibraltar rock,) that stands on the east side of the Nepean river,
at the first rise of the plateau from the central plain of the Cumberland basin, near Mulgoa. On the opposite or western side of
the river, the country rose very gradually for many miles, forming
a gently sloping plane, over the farthest edge of which (itself of
very considerable altitude), were seen some of the loftier peaks of
the central mountain chain. This sloping plateau had no remarkable eminence rising from it, but was everywhere traversed by an
infinite ramification of precipitous gulleys and ravines. Its surface indeed seemed about equally divided between these deep
valleys and their separating walls of rock. To traverse such a
country would be impossible except by either adhering to the top
of one of the ridges, and patiently tracing out its labyrinthine fold-
ings, always endeavouring to find its central axis, and never to
diverge into one of its branches, inasmuch as each branch would

probably split up into another maze-like net-work of ridges; or
else by descending at some broken point into one of the valleys and
undertaking the same almost hopeless task of tracing out one
onward track among an infinite number of ravines perpetually sub-
dividing. Higher up the sandstone plateau, when it attains an altitude of nearly 3000 feet above the sea, some of these valleys attain
dimensions of still greater grandeur.* They are however equally
numerous wherever the formation occurs, even if it be near or
below the level of the sea. Sydney harbour and Broken Bay are
instances of this, their many long winding arms and multiplicity
of small coves and bays, often bordered by precipitous cliffs, are
nothing but the gulleys and ravines of this formation, which in

---

* The following striking description of one of these valleys by
Mr. Darwin will give a good idea of their singular aspect and
character :—
" In the middle of the day we baited our horses at a little inn
called the Weatherboard. The country here is elevated 2800 feet
above the sea. About a mile and a half from this place, there is
a view exceedingly well worth visiting. By following down a little
valley and its tiny rill of water, an immense gulf is unexpectedly
seen through the trees which border the pathway, at the depth of
perhaps 1500 feet. Walking on a few yards one stands on the
brink of a vast precipice, and below is the grand bay or gulf (for
I know not what other name to give it), thickly covered with
forest. The point of view is situated as if at the head of a bay,
the line of cliff diverging on each side, and shewing headland
behind headland, as on a bold sea coast. These cliffs are composed of horizontal strata of whitish sandstone ; and so absolutely
vertical are they, that in many places, a person standing on the
edge, and throwing down a stone, can see it strike the trees in the
abyss below; so unbroken is the line, that it is said in order to
reach the foot of the waterfall, formed by this little stream, it is
necessary to go a distance of sixteen miles round. About five
miles distant in front, another line of cliff extends, which thus
appears completely to encircle the valley ; and hence the name of
bay is justified, as applied to this grand amphitheatrical depression. If we imagine a winding harbour, with its deep water
surrounded by bold cliff-like shores, laid dry, and a forest sprung
up on its sandy bottom, we should then have the appearance and
structure here exhibited. This kind of view was to me quite novel,
and extremely magnificent."

those localities are at a sufficiently low level to admit the waters of the sea.*

At Port Stephens, a little north of Hunter's River, are some great hills, and a wide district of porphyry, which has often a syenitic character. Sandstones of the palæozoic formation sometimes are seen leaning on the porphyry at very high angles, and containing in some instances pebbles of the porphyry or some very similar rock; but in others they are traversed by dykes and intrusive masses of a rock almost equally similar. The relative age therefore of the porphyry can only be discovered by a much more detailed survey than I was able to make of the district.

North of Port Stephens, our information becomes very scanty. From "Hodgkinson's Australia," however, we learn that there is sandstone and limestone (probably palæozoic) in the hills between the rivers Hastings and M'Leay. That there is granite, trap, clay, slate, and limestone, at various parts of the M'Leay river, and also in the hills which form part of a great lateral spur that projects from the main chain between the rivers M'Leay and Clarence. We have also in the same work indications of palæozoic rocks between those places and Moreton Bay, and in the latter district we learn

* See "Strzelecki's Physical Description of New South Wales and Van Diemen's Land, pp. 87-92." "Sir T. Mitchell's Travels in Australia." "Rev. W. B. Clarke, in Journal of Geological Society, vol. for 1848, pp. 60-66." And Jukes's "Notes on Palæozoic formations of New South Wales, and Van Diemen's Land," Journal of Geological Society, 1847, pp. 241-249."

the existence of coal and limestone (probably palæozoic) on the banks of the river Brisbane.*

Leichhardt† mentions the existence of coal on the
west side of the mountain chain at Darling Downs
and at Charley's creek in lat. 26° 40', and also near
Wide Bay on the coast.

D.—*The N.E. Coast.*

About Sandy Cape and Harvey Bay we have
no account of the rocks on the coast, and Leichhardt merely mentions the occurrence of a clayey
psammite on his track in the interior about those
latitudes. At Rodd's Bay and Port Curtis, Capt.
King mentions granite and sandstone as the principal rocks, and at Keppel Bay greyish slate,
quartz rock and various granitic rocks.‡

At Port Bowen in lat. 22° 30', my own observations on the N.E. coast commenced, and were
continued at intervals up to Cape York and Endeavour Straits. All the hills round Port Bowen
were composed of a heavy dull red porphyry like
that of Port Stephens. About Cape Townsend the
cliffs consisted of mica slate, traversed here and
there by large granitic veins. The group of islands
north of Cape Townshend, called the Percy and
Northumberland Islands, are composed partly of
quartz rock, partly of flinty slate, and compact feld-

* "Australia from Port Macquarrie to Moreton Bay," by
Clement Hodgkinson.

† "Journal of an overland Expedition from Moreton Bay to
Port Essington, by Dr. Ludwig Leichhardt," p. 17.

‡ See Captain King's Voyages of Discovery around Australia.

spar, together with some hard sandstone with quartz pebbles, while some of the low flat islands are made of a soft red sandstone.

West Hill (lat..21° 50′) is composed of basalt: the country around partly of sandstone and partly of clay slate and siliceous slate, with some soft recent sandstone near the beach. Similar rocks were found about the bays north of Cape Palmerston. Cape Hillsborough is composed of a base of igneous rock, capped by a mass of stratified rock, consisting of hard and soft sandstones, dipping S.W. at 15°. In the country at the back of it were small conical hills of columnar basalt. From a small island, a few miles north of Cape Hillsborough, I got a fragment of fossil wood, the only fossil I found on the N.E. coast. The island and apparently the coast behind it consisted of a hard whitish sandstone, weathering black and red, beds thick and much jointed. Around Whitsunday passage, the rocks were mostly igneous, either an amorphous basalt, a greenstone or a porphyry; others, however, had the appearance of highly altered stratified rocks, with the marks of stratification nearly obliterated. At Cape Upstart we came on true granite very largely developed, to the exclusion of all other rocks whatever.

The rocks mentioned by Leichhardt as found in the interior, between the latitudes of Sandy Cape and Cape Upstart, and at a distance of about 100 or 150 miles from the sea are the following; sand-

stone and fossil wood about Lynd's range (p. 40);
in lat. 25° 10′ five parallel ranges striking north and
south, and one running east and west are composed of
sandstone with impressions of calamites* (pp. 53-4);
psammite on Zamia creek (p. 61); much basalt
about Expedition range (p. 73); fossil fern leaves
and pebbles of coal on the M'Kenzie River (p. 105);
sandstone and coal shale with beds of coal ditto,
(p. 108.) He then mentions sandstone as extending
for a considerable distance interrupted by basalt, do-
mite and other igneous rocks, (pp. 119 to 152.) He
describes the country north of Isaac's River as a basaltic table-land, along the edge of which a series of
domitic cones run S.E. and N.W. to the coast, and
from them a nearly horizontal sandstone, locally
disturbed by basalt dipping towards the coast,
(p. 153.) He then mentions "flint-rock," and
pegmatite about the head of the Suttor River, and
after that porphyry, gneiss, and talc schist with
psammite, sandstone and quartzite, (p. 189.) He
then comes on sienite and granite but with pudding-
stone and limestone again recurring, (p. 191-2.)
At the juncture of Cape and Suttor Rivers he enters
on a widely spread granitic region (p. 195.)

This granite seems to be the extension of that
seen on the coast about Capes Upstart and Cleveland,
as Mounts Abbot and Elliott some distance in the
interior were also evidently composed of granite.

* Vegetable impressions doubtless, but whether calamites or
not is doubtful.

On the coast indeed from Cape Upstart to Cape Melville all the points and islands on which I landed were almost entirely granite. At Endeavour River, in the hills on the south side a granular quartz rock with lines of lamination was seen resting on or abutting against the granite. Cape Melville itself and the adjacent country is composed of one huge mass of granite, jointed into very large blocks, many of which have fallen from the precipices or slid down the slopes of the hills and now lie in the wildest confusion.

Turning again to Leichhardt's account of the interior we find that after he passed the granitic district of the Cape Upstart country, and the lower part of the Burdekin River (p. 210), he came upon rough basalt, hills of horizontal limestone, sienite with bent and uplifted strata of limestone, containing many fossils and then to fields of basalt again, (p. 211, &c.) In lat. 19° 30' we hear of baked sandstone and limestone full of corals.* North of this we find mention made of feldspathic porphyry, sandstone, and conglomerate (p. 222), porphyry and psammite associated with talc schist (224), and then of more talc schist, porphyry, and basalt, with gneiss-like hills, and broad sheets of basalt in the valley of Lagoons (pp. 226-8, and 240), north of the valley of Lagoons is a basaltic table land,

* I believe from the notes of the Rev. W. B. Clarke, who examined and described some of the specimens, that the fossils mentioned here are paleozoic.

with granite to the eastward of the river which forms the division between them (p. 244), after which scarcely any other rock but granite is mentioned till they began to descend the river Lynd, north of the Kirchner range in lat 17° 20′, north of that point nothing but sandstone is mentioned or psammite (large grains of quartz mixed with whitish red or yellow clay), till they got down to the sandy plains of the Gulf of Carpentaria.

The only points of the coast of the main land which I visited north of Cape Melville were Cape Direction, which was composed of granite fronted by sand, and Cape Grenville, where the only rock seen was a brown quartz rock with dark compact feldspar. The islands near the latter Cape, were either sand banks on coral reefs or rocky islets, like Sunday Island, and Sir C. Hardy's Islands, which were composed either of compact feldspar, of quartz rock, or of brown feldspathic and siliceous slates and schists. Around Cape York as well as in all the adjacent islands, Mount Adolphus, the Possession Islands, Prince of Wales's Islands, and Booby Island, no other rock was seen than a heavy compact dark coloured porphyry or greenstone.

E.—*Torres Straits.*

North of Cape York a narrowing chain of high rocky islands stretches right across Torres Straits up to the coast of New Guinea. Of these I believe Mount Augustus and Mount Ernest to be granite though I did not land on them, Saddle Island was porphyritic greenstone, Turtle-backed Island was nearly all granite except its N.W. corner, which was a brown and liver coloured laminated quartz rock. Cap Island* is composed of sienite, and from its shape I concluded Mount Cornwallis was also granitic. Around these islands in this north and south central band of Torres Straits, the depth of the sea is very uniform being from eight to ten fathoms, the bottom being nearly everywhere mud. On the N.W. side of it are some shoals said to be coral reefs, rising apparently from no great depth of water. The whole of the eastern part however of Torres Straits is occupied by true coral reefs forming the northern part of the great barrier reefs. The islands here with three exceptions are all flat coral islets, rising not more than twenty or thirty feet above high water mark, the water round them gradually deepening to the eastward. The exceptions are the groups of the Murray Islands, Erroob or Darnley Island, and some rocks on Caedha or Bramble Key. The rocks in these three localities are volcanic, consisting partly of sandstone and conglomerate made of pebbles of lava and coral limestone, with some beds of finer tuff, and partly of large masses of dark heavy hornblendic lava. The eruption of these volcanic rocks though probably of comparatively modern origin, geologically speak-

* Mr. Darwin in his "Coral reefs" supposes Cap Island to be volcanic from an erroneous description quoted in the introduction to Flinders' Voyage.

ing, must yet historically have been of ancient date, as no traces of any crater are apparent. From the occurrence of pebbles of coral limestone, they are almost certainly of subsequent origin to the commencement of the coral reefs here, but may yet date back into some tertiary period.[*]

The inspection of the " Chart of the Great Barrier reef and Torres Straits" will suggest to us some speculations worthy of note. First of all, the north and south central band of Torres Straits, or submarine continuation of the eastern chain, forms the western boundary of the Great Barrier reefs, just as much as does the dry land to the southward. No true independent coral reefs infringe on this band though they come close up to it. There are no wide spread coral reefs round its islets as there are round Darnley and Murray Islands, but only small fringing reefs clinging to the rocks. From this fact results two questions, 1. Are the " reefs seen by the Hormuzeer and Chesterfield" true coral reefs, or are they only banks of mud and sand with, perhaps, coral settled on them? 2. What are the islands called Duncan's Archipelago composed of? If not coral (and I believe they are not) good harbours with water 8 or 10 fathoms deep will probably be found among them.

The next point is the sudden ending of the Great Barrier reefs to the northward, and their reluctance

[*] See the Chart of the northern part of the Great Barrier reef, and " Voyage of the Fly," p. 204.

to approach that part of the coast of New Guinea composed of alluvial materials, and coloured pale green in the chart. This results from the well-known inability of the coral polyps to live in sea water charged with mud, or at all mingled with fresh water. Inasmuch, however, as the large reef of Warrior's Islet runs up to the New Guinea Coast west of Bristow Island, it would appear that no fresh water is discharged there, nor if the Hormuzeer and Chesterfield Shoals be really coral reefs, for a considerable distance to the westward. This inference is likewise strengthened by the submarine continuation of the eastern chain of Australia striking this part of the coast, since if it do not absolutely stretch into the interior of New Guinea in the shape of high land or mountains it will very likely have a tendency so to do, and thus raise the land above the average level and deflect the waters to the northward and eastward. The fact of Murray Island, Darnley Island, and Bramble Key being of volcanic origin, renders it rather probable that Aird's Hill is likewise volcanic. Now we know that along the north coast of New Guinea runs a lofty volcanic chain, from which these look like off-shoots, in which case there is probably a range of high land, or sufficiently high to cause a water-shed, proceeding from the north coast of New Guinea to tne south, somewhere about long. 146° or 147°.* If

* Very high land was actually seen in that direction from H.M.S. Fly, although a long way in the interior.

D

such be the case it would throw off water to the southward and westward. We have therefore on each side of the coast coloured "alluvial" in the accompanying chart, a probable rise of land, and an absence of the fresh water channels which are so abundant in that tract, which makes it probable that a great portion of the central region of New Guinea is drained by a river or rivers which disembogue on that "alluvial" coast. This would account for the vast body of fresh water we found there, not only keeping the multitude of arms and channels that intersect the land full of fresh water, but affecting the water of the sea to a very great degree, so as almost to render it drinkable at particular times of the tide, at a distance in some places of 8 or 10 miles from the shore.

Before quitting the eastern coast of Australia I must mention one singular circumstance, namely, the occurrence throughout its whole extent from Bass's Straits to Torres Straits of pebbles of pumice, strewed over the flats just behind the beaches to the height of a few feet above high water mark. These pebbles were always well rounded, varying in size from that of nuts to pieces as big as the fist, and are frequently found in considerable abundance. They have never been seen floating nor left on the actual beach by the wash of the sea, except in a few instances where they may have been washed down from above by the rains. It is difficult to conjecture their origin, but they may have proceeded from some old very violent eruption of the volcanoes of New Zealand, or of one or more of the foci of the volcanic band between that country and New Guinea. Their occurrence above high water mark does not necessarily involve the supposition of the elevation of the coast, as a wave of sufficient magnitude to float them into their present position may have been the result of one or more earthquakes which probably accompanied the exhibition of the volcanic violence by which they were originally produced.

In the southern part of New South Wales, indeed, and in Tasmania there are evidences of elevation of the land, not only in the tertiary rocks, now high above water, but in raised beaches and in beds and accumulations of sea shells beneath the present soil to very considerable altitudes; but all along the N.E. coast of Australia I looked in vain for any conclusive evidence of such a fact, and the existence of the Great Barrier reef off that coast would, if Mr. Darwin's theory of coral reefs be the true one, involve the supposition of great depression having taken place through a long period of geologically recent time. On this subject I have spoken more fully elsewhere.*

We have now to leave the eastern coast and proceed in our examination of the country towards the west.

Let me first state that the coast on the south side

* Voyage of the Fly, vol. i. chap. 13.

of Endeavour Straits, west of the hilly range that ends in Cape York and Peaked Hill was an absolute flat, scarcely raised above the sea, composed apparently of loose or slightly coherent sand. All observers state that the whole eastern and southern coast of the Gulf of Carpentaria has the same or a similar character.

From this region, where we thus get a peep at the nature of the country on the western side of the eastern coast range, we must now pass at once to

II. Australia Felix or the Port Phillip District.

I will first of all give my own notes of what I saw in this district.

My stay there was limited to 12 days, and my explorations were confined to a few miles, round Melbourne on one side of Port Phillip and round Geelong on the other. The town of Melbourne stands upon sandstone, a good deal resembling the Sydney sandstone in lithological characters. The beds are frequently inclined at angles of 20° or 30° dipping in different directions. No fossils were seen in them. A few miles south of Melbourne, at a place called St. Kilda, is a hard white quartzose sandstone, thin and regularly bedded. It is exposed in the cliffs, and on the beach at low water, and at one place I walked for full a quarter of a mile over the edges of beds dipping regularly to the N.N.W. at an angle of 45°, so that the thickness there must be at least from 900 to 1000 feet. No fossils were seen in these rocks either, but from mineral character, from their aggregate thickness, and from their highly inclined positions, I venture to consider them as belonging to the palæozoic formation. This supposition is supported by the fact of coal being found in the neighbouring district of Western Port, in rocks, as they were described to me, of precisely similar character. Upon these highly inclined strata repose at various places beds of a tertiary formation. These were well shewn a little south of St. Kilda, and at Brighton a mile or two beyond. The tertiary rocks here were composed of a brown ferruginous sandstone, commonly soft and friable, but often containing and sometimes passing wholly into a hard dark mass of ironstone more or less metallic looking and always concretionary. In other places the sandstone instead of brown is white, or red, or mottled with those two colours in various portions. These characteristics precisely resemble those of the rocks about Port Essington, on the north coast of Australia. The rocks of Port Phillip, however, often contain lines of fossil shells, mostly in the state of casts, these were chiefly referrible to the cultellus, amphidesma, donax, tellina, nucula and littorina, there were also numerous specimens of a small spatangus, and leaves of a dicotyledonous tree, very like eucalyptus in appearance. The thickness of these rocks exposed in these cliffs of " Brighton" was about 40 or 50 feet, the beds being horizontal.

The only other rock seen about Melbourne was a heavy hornblendic trap or lava, sometimes very heavy and compact, but often cellular and even scoriaceous, with all the appearance of recent subaërial lava. This rock is very largely diffused over the country as far as 12 miles N.E. of Melbourne, often occupying the valleys and lower grounds, and making a rough uneven surface, bristling with lava blocks. The higher grounds appeared to me to be usually of white sandstone. In some instances this lava seems to have flowed down the existing valleys of the country. The valley which comes down from the Moonee ponds into the salt lagoon near Melbourne, has on each side of it a high terrace of sandstone. Between these banks the bottom of the valley is entirely occupied by the trap or lava, in which has been excavated the present bed of the river, or whatever water it is that comes down the valley in seasons of flood.*

---

* In speaking of rivers and lakes in Australia, we must always divest ourselves of our European prejudices, of taking the presence

The general structure of the country around Geelong, is the same as that around Melbourne, an old formation believed to be palæozoic, a modern tertiary one, and rocks certainly igneous and apparently volcanic. The Barraboul hills, a few miles N.W. of Geelong, consist of a light olive brown or greenish sandstone, sometimes micaceous, sometimes grey and hard internally, at others a good freestone, very thick bedded, occasionally containing a line or two of small quartz pebbles. Reed-like vegetable impressions like coal plants were seen in it, and on the N.W. side coal one or two inches thick, was said to have been passed through in sinking a well. It dips in various directions between south and east on the Geelong side of the hills, at angles varying from 12° to 25°. It is traversed by one or two considerable faults perceptible in some of the cliffs ; and at one place grey grit and indurated shale were seen beneath the sandstone. These I believed to be palæozoic rocks. Between the Barraboul hills and Geelong, there were some flats and gently rising grounds occupied entirely by the rough scoriaceous lava or trap, (locally called ironstone). There were likewise some gentle hills of tertiary rocks. At one place these were exposed in an excavation for a lime kiln, and they were found there to be a coarse yellow rubbly limestone, very like some of the oolitic limestones of the Cotteswold hills, full of fossil shells, principally pectens, but others of the same species as those in the tertiary sandstone, near Melbourne ; there were also the same spatangi. This coarse limestone was not burnt in the kiln, but a white earthy very pure limestone, which was found as a shell or coating over the whole hill, not more than 6 inches in thickness, and immediately beneath the vegetable soil.

---

of water as a necessary consequence. The rivers and lakes of that country, only contain water as the exception, not as the rule. They are the places occupied by the drainage of the country, whenever there is any moisture to drain, that is after heavy and long continued rains.

In the cliffs of Port Phillip, on the west of Geelong, is seen about 40 feet of brown and yellow sand, with layers of soft sandstone, containing no fossils. On the east of the town, cliffs from 24 to 50 feet high, extend for about a mile, in which at half a mile distant are seen beds of a white concretionary limestone, sometimes tufaceous, and often a regular travertine. The bedding was at first irregular, but it passes towards the east into a smooth compact cherty limestone, with regular beds dipping W.S.W. at 5°. These beds contained small shells of the genera helix and planorbis, and the limestone was like a fresh-water limestone. It is not known anywhere except in these headlands, between Geelong and Corio. This limestone is therefore the monument of some old fresh water lake that must have existed formerly in this tract, when the physical geography of the neighbourhood must have differed considerably from its present state. The town of Geelong itself, stands on the lava, and the ridge between the port and the river Barwon, as well as the bed of that river, for about one mile above, and two miles below South Geelong, seems to consist entirely of the same rock.

Both in the neighbourhood of Geelong and Melbourne, recent shells may be seen, either scattered over the ground, mixed with the soil, or accumulated in ridges showing a comparatively recent elevation of the land. These sea shells are believed by the colonists to have been brought and left by the natives, and small heaps that have been so brought may be seen often enough. The shells I mean, however, are all considerably and equally decomposed, shewing that they were deposited at the same time or nearly so, and that that time was an ancient one. Besides which, they exist in numbers greater than could have been brought by the natives, unless the whole population of Australia had been employed in carrying them for centuries.

I now proceed to lay before the reader what has been described by others of the structure of the Port Phillip district, or the country south of the

Murrumbidgee and Murray rivers. Station Peak about fifteen miles north of Geelong, is, I believe, composed of granite. Mount Macedon still farther north, is said by Sir T. Mitchell to be sienite. N.N.W. of that, is Mount Byng, near latitude 37° and longitude 144°, which the same traveller describes as granite, as well as the range of Mount Hope and Mount Pyramid, sixty miles still farther north. He mentions clay slate on the banks of the river Campaspe in that neighbourhood, but he then speaks of nothing but granite with occasionally clay slate and quartz rock in the country to the N.E., till we reach the upper part of the Murrumbidgee River. For instance,—he describes the banks of the Barnard River as granite partly covered by trap, beyond which is clay slate and conglomerate, granite at the Swampy River, granite hills beyond the Ovens River,—granite at Mount Ochtertyre, on the banks of the Murray, — first range beyond the Murray granite. Mount Trafalgar, sienite, with clay slate and quartz rock, Burnett's range, granite. He then states, that in all the country where the Jugion Creek, the Dumott River, and the Coodradigbee Rivers fall into the Murrumbidgee, the higher country is composed of granite, the lower of limestone full of fossils, which he calls Silurian, and mentions favosites, stromatopora, heliopora pyriformis, and stems of crinoidea. From his descriptions, these palæozoic rocks with sandstone, " like that of the coal measures," seems to stretch across the main chain, hereabouts, through Yass Plains and by Lake George to the Shoalhaven River, as mentioned before (p. 18), the granite on which they repose, being, however, here and there exhibited at the surface.

Returning to Mount Byng and following Sir Thomas back to the eastward, we read of hills of vesicular lava, till we approach Mount Cole, and a north and south range called the Pyrenees, which consists entirely of granite, as does the country around Weelbong sixty miles to the northward of Mount Cole. East of Mount Cole, vesicular lava and basalt again occur up to the foot of the Grampians. The Grampians are described as a bold range of hills, 4000 feet in height, running north and south for about fifty miles, and composed from top to bottom of thick bedded ferruginous sandstone. Trap appears to the N.E. of them, while to the west in the upper part of the Glenelg River much granite was seen—and near the Wando River, gneiss, granular feldspar, and fine grained sandstone.

All the lower part of the Glenelg River, Sir Thomas describes as traversing a country evidently composed of the tertiary formation. He speaks of " a thin stratum of shelly limestone bearing some resemblance to some of the oolitic limestones of England," with " irregular concretions of ironstone containing grains of quartz, some glazed externally by a thin coating of hœmatite." Near Fort O' Hare, he describes "limestone rock with fossil oyster shells, at the foot of the cliffs and on the summit of the hills;" and below it, he mentions "picturesque cliffs of limestone varying very much in hardness, containing corallines, spatangi, echini, oysters, pectens, and foraminiferæ," "a thin stratum of compact chert with the same fossils,"—and in another locality, he describes the rocks as containing a nucula and portions of lucina, and turritella or melania. This tertiary formation stretches to Portland Bay, being, however, frequently interrupted by trap and vesicular lava: hills of lava often occur, and one at least, Mount Napier, is described as still exhibiting a perfect circular crater on its rugged summit, with thin walls of lava.

From the description I received I have every reason to believe that this tertiary formation dotted with lava hills, probably only extinct volcanoes, spreads over all the comparatively low country between Sir T. Mitchell's track above described and the sea up to the neighbourhood of Port Phillip. I was told at Geelong that Lakes Colac and Carangamite, about 70 miles west of Geelong, had cliffs of clay and sand abounding in fossil bones, some of which from their great size were supposed by those who had seen them to be the bones of whales.

III. Country from Australia Felix to the Darling River.

Of this great expanse of country our notices are

but few and scanty. We possess absolutely no information as to the structure of the district round the rivers Murray and Murrumbidgee, after their leaving the neighbourhood of the eastern chain till their junction, nor for some distance below it, except the general intimation that the country is low, devoid of any striking features, and characterized by barrenness rather than fertility. We meet with some better details of the district of the Lachlan and the Macquarrie, both in the old travels of Mr. Oxley, and the more modern ones of Sir T. Mitchell.

The latter in leaving Mount Victoria, the crest of the eastern chain near the latitude of Sydney, passes over granite in the vale of Clwyd, and over porphyry in approaching Bathurst. Between Bathurst and the Conobolas, he mentions trap and limestone; about Buree, and for some miles to the N.W. nothing but granite; near Mount Juson he finds trap, but Mr. Oxley describes Harvey's range to the north of that as consisting of granite. North of the Goobung River, Sir T. Mitchell describes schistose rocks dipping to the east at 60°, while around Marga S.W. of Buree, he mentions purple clay slate and a ridge of granite. Mounts Amyot, Cunningham, and Allan, are said to be ferruginous sandstone, like that afterwards to be described on the Darling River. We then hear of chlorite slate at Hard Hill, and of quartz, clay slate and ferruginous sandstone at the Kalingalungaguy River, where the general strike of the rocks is said to be N.N.E., dipping sometimes westerly, sometimes easterly at high angles. At Tarratta we again meet with granite, and in the plains beyond with porphyry, while near Regent's Lake, Sir T. Mitchell mentions trap and calcareous conglomerate like the tufa of caverns. We then come to Peel's Range on the south of the river Lachlan, which Mr. Oxley says is sandstone resting on granite, and to other hills on the north of the river, of which Mount Garnard is one, among which Sir T. Mitchell mentions the occurrence of hornstone (granular feldspar)? trap and quartz rock. The most westerly hills of this range, which runs north and south, are however, all ferruginous sandstone, and from them the view is uninterrupted to the west, and extends over the apparently boundless plains which spread round the junction of the rivers Murray, Murrumbidgee, and Lachlan, and stretch thence to the banks of the Darling. We have now only to follow Captain Sturt and Sir T. Mitchell, in their course down the Macquarrie, the Bogan, and the Darling. On the former river Captain Sturt describes decomposed mica slate, feldspar and schorl rock as occurring at the falls a little north of 32°, and basalt at Mount Harris and Mount Foster in about 33°. On the Bogan, Sir T. Mitchell met with mica schist, and quartz, whinstone, granite and porphyry, near where Mr. Cunningham was killed about lat. 32° 10′ and long.

45

147°. At Mount Hopeless, near lat. 31°, he mentions quartz rock, and quartz and feldspar. North of New Year's Range, Captain Sturt and Sir T. Mitchell concur in describing the bed of the Bogan* as granite, but beyond that neither authority mentions any other rock than ferruginous sandstone. Sir T. Mitchell indeed describes the general features of the Bogan as ferruginous sandstone resting on granite, but beyond its junction with the Darling no granite is seen, and the bed of the Darling as well as the neighbouring eminences of New Year's Range, Oxley's Table land, the D'Urban group, Dunlop's range, Mount M'Pherson, Greenhough's group, and Mount Murchison, the latter of which is in long. 143°, are all described as ferruginous sandstone, indurated sandstone or quartz rock. Mention is made of gravel of quartz pebbles, near the river and even on the summits of the hills, and a clay band three or four miles wide on each side of the river is also described. The surrounding country is everywhere described as a barren plain, with the few scattered eminences mentioned above, which are apparently of no great height or importance, rising from it at intervals.

IV. SOUTH AUSTRALIA.

The most striking and important feature in South Australia, is the hilly chain which runs from Kangaroo Island and Cape Jervis, at first about N.N.E.,

* New Year's Creek of Sturt.

but from the latitude of Adelaide about north till it terminates near lat. 29° and 30° in the singular horse-shoe shaped depression of Lake Torrens. Its length is thus about 400 geographical miles. None of its peaks ever much exceed 2000* feet in height, but it is broken into many ranges, often having the peaked and serrated character of a mountain chain in miniature. About Cape Jervis I was informed the rocks were chiefly mica slate and gneiss; in the hills behind Adelaide, I found much chloritic slate and similar rocks, but north of that for about 50 miles, or as far as the mines of Kapunda, the rocks of this chain seemed to be principally clay slate.

Wherever I saw the dip of the rocks shewn, it was to the S.E. at pretty high angles. This would put the clay slate under the gneiss, mica, and chlorite slate, but my observations were so few that the inference is not to be depended on. The only observation on the cleavage I made was about five miles N. of Gawler Town, where both the cleavage and the bedding of the clay slate dipped S.E., the cleavage at a higher angle than the bedding. Many large white quartz veins traversed the clay slates, all the peaked eminences being apparently composed of white quartz.† Around Kapunda, though in the heart of the chain, the country was only gently undulating. The mines are about a quarter of a mile N. of the bed of the River Light. Though nothing but hard clay slate was to be seen in the adjacent country, the rocks near the mines are all soft,—generally a blue, white, or pinkish

---

* Mount Lofty, near Adelaide, is said to be 2,700 feet high.

† We were only five days in South Australia, and my own observations were limited to a day round Adelaide, and a ride of three days to Kapunda and back. Whenever, therefore, I speak from actual observation, its cursory nature must be allowed for.

arenaceous claystone. The galleries of the mines are dug with pickaxe and spade, and the rock can be easily cut in all directions with the knife. Some of it hardens rapidly when exposed to the air, so that when dug out and cut into shape it forms in a few days a very good building stone. In the galleries of the mines, the rock, which could only be called an argillaceous sandstone, had a "stripe," or appearance of bedding similar to that so often seen in clay slate. The general dip was W. 10°, S. at about 15°; but the rocks were said to have a quaquaversal dip of about that amount in this locality. There were eight lodes of copper ore, running parallel to each other, in the direction of N. by E. and S. by W., seven of them nearly perpendicular, but one in the plane of the bedding. The ore consisted of blue and green carbonates, and black oxide, the latter most abundant in the deeper parts of the mine. The rock is said to get harder as they descend. The greatest depth they had attained in 1845 was 16 fathoms, or 96 feet, at which depth they got some water, very clear, but very salt, and aluminous to the taste.*

About forty miles north of Kapunda is the celebrated Burra Burra mine, which I had not time to visit; but which was described to me by Mr. Kingston, of Adelaide, who was concerned in its management. He said the general rock of the country was clay slate there also, but the copper ore, instead of a regular system of lodes, was found in one large mass, occupying the surface of the ground for many hundred square yards. Holes had been sunk into it, and in one place, at the depth of 26 feet, they had pierced through the ore into a similar soft rock to that which forms the "country" (as miners term it) at Kapunda. Other openings had been blasted and quarried to a greater depth in the copper ore without reaching its boundary, nor were its limits at the surface exactly known at that time. Some distance from the working, a stout quartz vein, running about N.N.E.

---

* All the water in the neighbourhood had an aluminous taste, which those accustomed to it liked, but which was very disagreeable to strangers.

and S.S.W., seemed to form the limit of the copper ore in one direction. The most singular thing respecting these mines appears to me to be the softness of the rock in which they lie. I believe this rock to be part of the surrounding clay slate formation, but either it can never have been metamorphosed into clay slate, or some agency has remetamorphosed it back again into its original condition, or something like it. Can that agency be in any way connected with the production of the mineral veins and masses? I regret that my time was too limited to allow of my making further observations on these interesting points. My search for organic remains was unsuccessful, but it was necessarily very imperfect.

Of the remainder of this hilly chain of South Australia, we only get a few other geological notices from Mr. Eyre's travels. He describes Flinders' Range, near Mount Arden, as consisting of clay slate, argillaceous stone and quartz; Mount Deception (lat. 30° 48′) as 3000 feet high, and composed of micaceous slate; and the rocks N.E. of Mount Eyre as quartz and ironstone. Mr. Kingston of Adelaide informed me that he knew of no true granite in this South Australian chain, but Mr. Dutton, in his "South Australia and its Mines," page 259, mentions granite as occurring in the beds of the rivers and forming the peaks of the hills of the Barossa range. Its appearance is at all events rare.

We have now to speak of the country intermediate between the South Australian chain, and the region of Australia Felix before described. We saw that all the lower part of the river Glenelg was occupied

by a large tertiary formation, containing beds of
limestone, full of fossil shells of existing genera.
From this point to the banks of the river Murray,
we have no published accounts of the country, but
I have always heard it spoken of as low, nearly
flat, and consisting generally of barren plains. A
little west of the Glenelg are the "biscuit plains,"
so called because they are covered with thickish,
flat, round stones, of the shape and appearance of a
sea-biscuit. These are generally smooth above,
and have a curious cellular structure below, but
whether they are the remains of anything organic,
or merely concretions or incrustations, I have not
been able to ascertain. On arriving at the banks of
the Murray below the junction of the Darling, and
thence to Lake Alexandrina, we find the tertiary
formation alone exposed. Captain Sturt describes
this formation both in his first travels and in those
lately published, figuring many of the fossils in the
Appendix to the first.

In his last publication, "Sturt's Expedition into
Central Australia," he describes the "fossil formation," as he calls it, as rising to the west on the
bank of the Murray near Lake Bonny (long. 140°
30') to a height of 250 feet, as having thence a
regularly horizontal stratification, with a gently
undulating surface round the great bend of the
Murray, till on approaching Lake Alexandrina, it
suddenly dips and disappears. He says, " its lower
part is entirely composed of turritellæ, but every

E

description of shell with bones and teeth of sharks
and other animals have been subsequently found in
the upper part of the beds, and the summits are in
many places covered with oyster shells, so little
changed by time as to appear as if they had only
just been thrown in a heap on the ground they
occupy," (p. 29). On comparing this account with
Sir T. Mitchell's description of the Glenelg, and
knowing there are level plains all the way between,
we can scarcely hesitate to conclude that these two
localities only exhibit sections of the same wide-
spread formation. It is worthy of remark, that in
descending the Murray, in his former travels,
Captain Sturt speaks of "ironstone" as the first
rock seen on approaching the tertiary formation.
He also describes in his last book, numerous sand
dunes on the face of the country which is composed
of the fossil formation, near the Murray, and says,
near Mount Misery and Rufus River (long. 141° and
thereabouts) the Murray is flanked by high level
plains on both sides, that the cliffs are 100 or 120
feet high, composed of clay and sand, the faces of
which are deeply and delicately grooved, so as to
present an appearance of fretwork (p. 89), and that
near the great bend of the Murray, (in long. 140°,
45'), the hills are of a yellow and white colour, the
rock being a soft and pliable sandstone (p. 88).
These rocks are evidently all either of the tertiary
formation, or are still more recent accumulations.
In describing a traverse he made from this spot,

(called Moorundee) to Adelaide, Captain Sturt
(p. 32) says, " I turned from the river to the westward, along the summit of the fossil formation,
which at a distance of a few miles was succeeded by
sandstone, and this rock again as we gained the
hills by a fine slate, and this again, as we crossed
the Mount Barker and Mount Lofty ranges by a
succession of igneous rocks. On descending to the
plains of Adelaide, I again crossed sandstone and
to my surprise discovered that the city of Adelaide
stood on the same kind of fossil formation I had
left behind me on the banks of the Murray."

Here I can again speak from personal observation.
About ten miles south of Adelaide, the flanks of the
chain of hills, which from Cape Jervis northward
had been washed by the sea, begin to recede from
the coast, and let in a low plain between them and
the water. This plain increases in breadth to the
northward, till about Adelaide it is ten miles wide,
and appears to be still wider towards the head of
the gulf of St. Vincent. It consists entirely of the
tertiary formation, patches of which are said to
occur here and there among the hills at a much
greater elevation, but these I did not see.

Around Adelaide the formation is well shewn in several quarries. In the higher parts of the town, immediately below the
surface, is found a light yellow concretionary limestone, forming
a thin bed, below which is a white clay, and below that, sand,
with other beds of stone. These latter are seen in a quarry at
the back of Government House, on the bank of the Torrens,

where, in a thickness of 25 feet, are seen several beds of a light-coloured, compact, hard, earthy limestone, separated from each other by layers of fine loose sand. Both in the sand and in the limestone are many marine shells, especially ostrea and pecten. Casts of shells, apparently venus or lucina and turritellæ, are also frequent, and some spatangi, of exactly the same species as those near Port Phillip and those I have seen from the banks of the Murray. I was struck with the perfect state in which the oysters were found in these quarries at Adelaide, and had mentioned in my notes their little altered condition, before I remembered that the same fact was mentioned by Sturt and Mitchell on the rivers Murray and Glenelg.

We will now briefly follow Captain Sturt in his last desperate attempt to penetrate into the central portion of Australia. Going up the Darling from the Murray he speaks merely of plains which we have seen reason to believe are all composed of the tertiary formation until we approach the Barrier or Stanley range, which rises in lat. 32°; there he speaks of the surface being covered with fragments of white quartz, which together with a conglomerate rock cropped out of the ground where it was more elevated (p. 151.)

In this range, the hills of which are in some cases nearly 2,000 feet high, he describes the occurrence of fine hepatic iron ore in great quantities, and speaks of soapstone, of gneiss, of granite, of coarse ferruginous sandstone, and a siliceous rock, p. 166. He afterwards mentions in this range trap, coarse grey granite, " immense blocks," and " huge masses" of granite, p. 175, &c. He then

comes to plains of *clay and sand,* covered by angular
fragments of quartz rock, ironstone, and granite,
(p. 180.) Mount Arrowsmith, in lat. 30° 5', the
southernmost hill of the next range called Grey's
Range, he describes as having indurated quartz
and pieces of gypsum, (p .219), a compact sandstone
with blocks of specular iron ore scattered over it,
highly magnetic, (p. 231.) He describes the hills
about Mount Poole, lat. 29° 35', as composed of compact white quartz, split into innumerable fragments
in the form of parallelograms (p. 240), and says the
same white quartz continued throughout the Grey
range, the neighbouring plains being covered with
its fragments, (p. 259.) Near the Depôt, however,
on Preservation Creek, he found large slabs of beautiful slates traversed by veins of quartz. We have
here then, evidently, two north and south ranges of
hills of moderate elevation, being together about
180 or 200 miles in length, and composed of old
igneous and metamorphic rocks. A little N.W.
of the extremity of Grey range, Captain Sturt
describes the Bawley Plains, the basis of which he
says is sandstone, on which rested thin layers of
clay, with ridges of loose sand reposing on the clay.
Thence he penetrated into the interior for about 400
miles, generally in a N.N.W. direction till he
reached lat. 24° 30', long. 137° 59'. For the whole
of this distance the country was a desert of the most
inhospitable character. The arid plains were covered
with bare ridges of drift sand, often 80 to 100 feet

in height, running in parallel lines to the horizon in each direction. A little scanty grass grew here and there, nourished at rare intervals by thunder showers. Permanent water there was none, and Captain Sturt made one expedition of 180 miles on the strength of a single shower of rain; the puddles of which had not wholly dried up on his return. The heat was intense, "lucifer matches," he told me, "took fire if dropped upon the sand, and the thermometer on one occasion reached 153° in the coolest spot they could find." In the centre of this terrible plain of sand ridges is a remarkable "stony desert" round lat. 26° 30', long. 139° 30'. This is the lowest and most depressed part of the interior; it is so thickly covered by stones as wholly to exclude vegetation; the stones are fragments of quartz which are said to be rounded by attrition. Its known dimensions are 35 miles by 80. He elsewhere speaks of its surface as a great expanse of solid dark ironstone on which the horse's hoof rang and left no impression.

To the N.W. of the Stony Desert, the sand ridges ran in lines bearing N.N.W. and S.S.E., and from his farthest point he saw no hope of their cessation. Between the Stony Desert and Grey's Range the sandy ridges ran in lines bearing N.N.E. and S.S.W. Between the north end of Grey's Range and the depression of Lake Torrens, which was struck in lat. 29° 10', long. 139° 50', the sand ridges curve from N.E. and S.W. (their direction near the hills) gradually round to N.W. and S.E., their direction near the Lake.* Captain Sturt discovered one oasis in this dreary desert of the interior, a river-like sheet of water which he called Cooper's Creek. This near latitude 27° 35′ extended in a nearly east and west direction for about eighty miles, ending each way in arid sandy plains. In seasons of flood it is supposed to run to the west towards the Stony Desert and Lake Torrens, which are believed to be the outlets of the drainage of the interior in those rare seasons when there is anything to drain. Mitchell's river Victoria, which rising near Mount Pluto, part of the great Eastern Chain of Australia, runs boldly into the interior, full of water as if the commencement of a river to carry fertility and navigation into the heart of the country, was traced by Mr. Kennedy, till it gradually dwindled away in small creeks and came to an end in lat. 26° 20′, long. 142° 20′, barely 100 miles from the eastern extremity of Cooper's Creek. It is probable, therefore, that if there were water enough these two would be the same great river, which in that case would run from the inner flank of the great eastern chain, parallel to the Darling and the Murray, rising

* In speaking of Lake Torrens the reader must be again cautioned not to call up to his mind's eye a sheet of water. It is merely a depression below the level of the surrounding country, in which mud and occasional pools are found, and which in a damp country would be permanently full of water.

in Mount Pluto, and disemboguing at the head of Spencer's Gulf.

Captain Sturt never mentions the occurrence of any fossils anywhere in this desert interior, but from our knowledge of the tracts of Australia already described and that round the Great Australian Bight, to be now entered on, we can hardly help looking at this vast plain of sand, resting on clay, on sandstone, and on ironstone, so flat and so little elevated above the sea, as having the tertiary formation for its substratum.

Of the remainder of South Australia, we have but little published information. Of the Yorke Peninsula, I can find no mention made. From the head of Spencer's Gulf to Port Lincoln, Mr. Eyre mentions no other rock than quartz, sandstone and sand. In crossing from the head of the same gulf to Streaky Bay, he passes first over a conglomerate and next over some quartzose grits, but the Gawler Range he describes as principally if not entirely formed of granite.

IV.—GREAT AUSTRALIAN BIGHT.

This district extends from the west side of the Port Lincoln peninsula to the neighbourhood of King George's Sound in the colony of Western Australia. The coast was surveyed by Captain Flinders, who describes its aspect from the sea, and it has been traversed by land only by Mr. Eyre,* in his most arduous and difficult over-

* Now Lieutenant-Governor of New Zealand.

land journey from South Australia to Western
Australia.* Near Coffin Bay he says the hills are
of oolitic limestone, with granite protruding through
it in one of the gorges. This oolitic limestone
belongs to the "fossil formation" which he subsequently mentions, and is no other than part of the
great tertiary formation previously described.
Mount Hope near Coffin Bay is granite, thence to
Streaky Bay nothing is seen but limestone and
sand, except Mount Hall near Streaky Bay, which
is granite. From Streaky Bay to Fowler's Bay it
is all tertiary, masses of marine shells were found
in it, and the country is said to be *crusted with
oolitic limestone.* From this point round the head of
the Great Australian Bight to the neighbourhood of
Cape Arid, a distance of about 600 miles, both Flinders and Eyre speak of an unbroken line of cliffs of
horizontally stratified rock varying in height from
three to five hundred feet. At the head of the
Bight Mr. Eyre gives us the following section.

1. Oolitic limestone in a crust probably of no
great thickness.

2. Hard concrete sand with pebbles and marine
shells.

3. Hard coarse grey limestone.

4. White substance like chalk.

* Journals of Expedition of Discovery into Central Australia,
and overland from Adelaide to King George's Sound, in the years
1840-1, by Edward John Eyre.

He frequently mentions the fossil shells, and describes them as of recent aspect. At the top of the cliffs the country is sometimes said to be covered with recent *freshwater* spiral shells, but as Mr. Eyre apparently does not pretend to be a conchologist, their freshwater origin may be doubted. He also speaks of places where the surface was strewed with broken flints like gun flints. These are probably the same as the angular pieces of quartz mentioned by Sturt as occurring in some parts of his central desert. Beyond (or to the westward of) the head of the Bight, the same formation was in the upper part brown, in the lower white, and in long. 129° the brown was found to be a coarse grey limestone with a few shells, very hard; the white a gritty chalk full of broken shells and marine productions, with horizontal layers of flint, six to eight inches thick. The lower white or chalk part was often worn away, and the upper brown overhung in crags with fallen fragments on the beach below.

Beyond long. 126° we get this section :

    1. An upper crust of oolitic limestone with shells.

    2. Coarse grey hard limestone.

    3. Alternate strata of white and yellow limestone in horizontal layers.

In approaching Mount Rugged, granite was again met with. Throughout this extent of 600 miles, Mr. Eyre always mentions the summit of the cliffs as stretching away into the interior in apparently boundless plains. He not only meets with no river, nor river course, but he fails in finding a single drop of fresh water on the surface of these plains, and he makes three successive journeys of seven or eight days a-piece with only so much water as he can carry with him. To obtain this he descends the cliffs to the beach at three several points where alone the descent appears to have been possible, namely near long. 128° 40′, near that of 126° 30′, and near that of 124° 10′.

In sailing along this immense extent of unbroken cliffs, an extent of which I can call to mind no other example in the world, Flinders was led to speculate on its being an elevated coral reef, and that behind them he would probably have found an inland sea. Captain Sturt, likewise, in his recent journey into the interior, seems always to have expected to come to some great inland water. This notion of an inland sea always seemed to me, since I have studied the subject, in the highest degree improbable, and we have now seen it in two instances dispelled. There can, I think, be little doubt that these great tertiary plains round the head of the Great Australian Bight stretch with almost unbroken continuity far into the very centre of the island, and in all probability join on to Sturt's great central desert of sand and ironstone.

To return, however, to Mr. Eyre's journey. He describes Mount Rugged and Cape Arid as being principally, if not entirely, granitic. Thence to Cape Le Grand he describes the extension of the tertiary formation with fossil shells, as having granitic knobs protruding here and there through it. At the Salt Lakes, east of Cape Le Grand, we still have the fossil formation at an altitude of 150 feet, but after that we have no farther mention of it. About long. 121°, however, we have sandstone and ironstone described as resting on the granite, and I am strongly disposed to look on this sandstone and ironstone as belonging to the same tertiary formation as the fossiliferous limestone, that the two kinds of rocks are either different parts of the same formation, or that they replace each other in the same geological horizon.

On arriving at East Mount Barren, Mr. Eyre speaks of micaceous slate, and we now enter the colony of Western Australia.

V.—Western Australia.

As I made an excursion of two or three weeks in Western Australia, I shall first of all give an account of what I myself saw of its structure.

In approaching Swan River you look over a gently undulating plain of some twenty miles in width, backed by a range of hills, called the Darling range, which runs north and south parallel to the coast, and has a mean height of 800 or 1,000 feet. On climbing these hills, however, you find, that instead of being a single ridge, as it appeared from the sea, it is merely the steeply sloping edge of a hilly district, which extends for many miles into the interior of the country. In traversing the plain from the sea, you first pass for about ten miles over a district of loose white sand, quite impassable for wheel carriages, but covered by the

usual forest of the country, and producing fruits and vegetables
in considerable quantities in the gardens of the colonists. It
consists partly of grains of quartz and partly of calcareous grains,
probably rolled fragments of shells and corals. In several places
it passes into the state of a rather soft friable limestone, sufficiently
firm to be used for building stone. In other places are seen
rising from the sand what appear to be trunks of fossil trees,
having not only the external form of trees, but much even that
resembles their internal structure. These occur throughout the
colony in places wherever this white sand is found, and they have
been frequently described at King George's Sound, where Mr.
Darwin believed them to be calcareous concretions formed in the
hollows left by decayed trees. In a little cliff near Fremantle,
however, near the entrance of the Swan, I saw some of these
dendritic masses fully exposed, and from their peculiar structure
and conformation I believed them to be nothing more than *stalac-
tites formed in the sand* by the percolation of rain water dissolving and taking up the carbonate of lime found in the sand, and
re-depositing it in fantastic forms wherever a predisposing cause
happened to determine it. I believe the limestone in these sands
likewise to be formed in the same way, as the bedding had frequently a rather highly inclined or contorted dip, evidently not
due to movements of elevation, but the result of their original
formation. In this case I suppose rain to have sank through the
sand, dissolving the carbonate of lime in its passage, till it at
length became saturated or could sink no farther, and that, as it
evaporated, the carbonate of lime was deposited in a crystalline
condition, binding up all the adjacent grains into a more or less
solid stone.

At the foot of Mount Eliza, near Perth, some beds of a close
red sandstone may just be seen beneath these white sands and
limestones, and a few miles further east, just on reaching the
ferry to Guildford, the upper white rock ends in a little escarpment, and the red sandstone crops from under it, and forms the
remainder of the country in the neighbourhood of the Swan
River up to the foot of the hills. This red sandstone forms a

much superior country to that of the white sand, the roads are hard and firm, and the ground far more fertile. The sandstone seemed often argillaceous or marly, and in some cases seemed to contain beds of hard grey gritstone and ferruginous concretions. In lithological character it resembled a good deal the new red sandstone of England. I never found any fossils in it, but it appeared every where to be horizontal and conformable to the white limestones and sands above it. These latter are certainly tertiary, as they in some places contained shells of the genera arca and venus apparently of species now existing on the coast, and certainly of very recent and unaltered aspect. I am inclined to believe, that the brown sandstones below, with ferruginous concretions, as they are certainly conformable to these tertiary beds, may in reality be a lower part of the same formation.

On approaching the foot of the hills on the road from Perth to York, we pass at once from the horizontal brown sandstone to a steep slope of granite, above this to another of greenstone, and then to granite again. The surface of this plateau of rock, here about 800 feet above the sea, was gently undulating, covered with one wide forest of a large eucalyptus, called here mahogany. For a few feet below the surface the rock was a singular concretionary ferruginous compound, which looked like a clay or sandstone that, being highly ferruginous, had formed itself into a mass of small balls and irregular concretions of a black oxide of iron or hæmatite. Below this "ironstone" (which is its name in the colony), wherever the rock was exposed, it appeared for many miles to be granite, or some granitic compound. At a place about twelve or fifteen miles from the edge of the hills, in the descent of a lateral valley of the Swan, I passed, in the space of a mile, over the following rocks :

1st. A thin capping of "ironstone," forming a line of small crags.

2nd. A band of coarse granite.

3rd. Fine rather soft chloritic schist, green with a silvery lustre.

4th. A very hard crystalline greenstone, passing here and there into sienite.

5th. Compact white quartz.

6th. A hard semi-columnar basalt, apparently a vein or dyke.

7th. A pinkish quartz rock.

8th. The same rock laminated and crystalline, and gradually becoming gneiss.

Thence, through Toodyoy, to a place called Bulgart, at that time the farthest inhabited spot towards the north-east in the colony, and about sixty miles from Perth, I found mention in my notes of greenstone, sienite, hornblendic and chloritic schist, gneiss and mica slate, with large intervening spaces of granite, with the details of which I shall not attempt to weary the reader. About three miles beyond Bulgart I was taken by Captain Sculley to the edge of the " sand plains," which were said to stretch to an unknown distance into the interior. These were plains of loose white sand covered with Banksia and other shrub-like trees and bushes, with bare sandy ridges rising at intervals. They much resembled the white sand plains of the coast, but the sand seemed more pure quartz, scarcely, if at all, calcareous, and the country was said to be hopelessly barren, utterly destitute of food except for a few kangaroos, and having only a few water-holes here and there known to the natives. From Toodyoy I rode up the Swan (here called the Avon) to York, and thence back by a different route to Guildford and Perth. In all the hill country I found the same rocks in similar variety with those previously mentioned.

I had carried a mountain barometer with me, and Mr. Roe, the Surveyor-General, was kind enough to make daily observations at Perth during my absence. By these means I was enabled to calculate roughly the heights above the sea of all the places I visited. I found I had never passed over any part much, if at all, exceeding 1,000 feet above the sea, but I also discovered that the bed of the Swan rose much more rapidly than I should have expected. From the coast nearly to the foot of the hills the river is affected by the tide. Where I first came down on it, on the road from Balup to Toodyoy, it was already 250 feet above the sea, having risen that height in a distance not exceeding twenty

or thirty miles ; near Northam it was 400, and at York full 500
feet above the sea, the distance from tidal water, even following
the windings of the river, certainly not exceeding one hundred
miles. The structure of the river among the hills was very peculiar ; just above York, near Mr. Landor's house, there was a pool
in the river bed 200 yards long, 30 yards wide, and 36 yards
deep, full of water to the brim, while just beyond each end
of this pool the river was dry, or had only a little winding,
trickling stream among the grass and pebbles. A succession
of these dry spaces and "water-holes," as they are called,
occur all down the river. That near Mr. Slade's house, below
Northam, was three-quarters of a mile long, thirty or forty
yards wide, very deep, and bank full of the most beautiful clear
water, with winding reaches like any other river. A stranger to
the country, coming suddenly upon it, would have concluded of
course he could have placed a boat on it, and gone down in it to
the sea, and yet it suddenly ended each way, with a steep square
termination, and the bed of the river beyond was merely a grassy
hollow, full of trees and bushes, a little above the level of the
water in these pools   I had, indeed, ridden across the bed of the
Swan lower down, without knowing it, and only became aware of
the fact by finding a long reach of water on my right, whereas I
knew the last I had passed had been on my left.   It seemed as if
somebody intending to deepen the bed of the river had set parties
to work at different places, and that these parties, after having
excavated different lengths of work to considerable depths, had
all suddenly left off, and the holes thus made had absorbed all the
water that formerly ran down the river.   I should have supposed
that this peculiar character had been the result of the rapid rise
of the bed of the Swan above-mentioned, but it is a character
which, from the accounts of travellers, and from all the brooks
and rivers I saw in the country, is more or less common to the
whole of the rivers of Australia, with the exception perhaps of the
Murray.   They all consist of a regular river bed and channel,
which is only used, as it were, *in extenso*, in wet weather, the
various holes remaining more or less full according to the dryness

of the season. How these holes were produced with their regular shape, their depth, their occupying for different distances the whole width of the river bed, and their sudden endings in each direction, was a difficulty I could not solve.

I have now to speak of what is known of the remainder of Western Australia. The steep edge of the hill country runs north and south for many miles north of Swan River, and south of it as far as the sea at Point d' Entrecasteaux; the rocks of which the hills are composed seem likewise to be much the same as in the neighbourhood of the Swan, but the hills must sometimes assume loftier and bolder features, as Mount William in lat. 33° is said to be 3000 feet high. Granitic and metamorphic rocks appear to form all the country from York to King George's Sound, but there the tertiary sand with concretionary limestone seems in places to rest upon them to a height of 400 feet above the sea at least. It is probable, also, that these tertiary accumulations occur at intervals all along the coast to the eastward till we reach the great tertiary district of the Australian Bight. The tertiary plain of Swan River likewise appears to stretch to the southward till it reaches the south coast, but south of the Swan I believe the white sand band gets narrower, and the plain is partly composed of clay and red sandstone. The projecting piece of land between Cape Leeuwin and Port Naturalist exhibits, I believe, granite on the coast beneath the tertiary rocks.

From the descriptions of Captain Grey and others

F

it appears certain that the tertiary sands and limestones extend northward from Swan River along the coast at least as far as Sharks Bay, and it appears equally certain that the hilly granitic and metamorphic country runs behind it for a great part at least of that distance.

From some recent explorations by the Messrs. Gregory and Lieut. Helpman, an account of which was sent about two years ago to the Geographical Society, we get some interesting information with regard to the country in about lat. 29°. From this it appears that much of the rock about Moresby's flat-topped range, Victoria Range, and Herschel Range is granitic, and on the coast both north and south of Champion Bay granite, and granite capped by sandstone and sandy limestone are mentioned, and it is almost certain that this sandy limestone is the same as that of Swan River. On the banks of the Irwin River, at a spot in lat. 28° 47′, long. 115° 30′, coal was found. Here it is said "two formations were visible, a lower mass of shales and sandstones dipping east,* and above in the cliffy sides of the valley red sandstone horizontally stratified." The coal had one bed, five feet, and another six feet in thickness. Some fossils from this locality were sent to the Geological Society,

* In a paper by Dr. Von Sommer on this locality, published in Journal of Geological Society, vol. v. p. 51, the dip is said to be W.N W. at 72°. This would seem to be the more probable dip, but the remainder of his observations are little to be depended on.

which were, I believe, mostly identical in species with those of the palæozoic formation of New South Wales. Eastward of where the coal was found was a widely spread granitic district, with trap, &c., and east of that again a "desert of sand." On approaching the hills from the coast it is said that the red sandstone was seen stretching on the adjacent hills horizontally for a long distance.* It appears, then, that in this latitude we have a repetition of the same facts as at Swan River, with the addition of a small patch of palæozoic rocks, amongst the granites, &c. of the hills, and the variation of the red sandstones spreading some way up on to the hills, instead of being confined to the flats of the coast.

Captain Grey† in his Journal describes Gairdner's Range (between Herschel Range and the coast) capped by ironstone for 14 miles, being the same, no doubt, as that seen by me on the hills near Swan River. He likewise mentions recent limestone with recent fossil shells and sandy dunes as forming the islands at the mouth of Sharks Bay (vol. i. p. 333 to 408), and speaks of the same limestone as forming the coast as far south as Gantheaume Bay, and

* The height of the hills hereabouts from the descriptions of Captains Grey and Stokes appears to be about the same as that of the hills near Swan River, namely, from 700 to 1000 feet.

† Journals of Expedition of Discovery in North-West and Western Australia during the years 1837, 8, 9, &c. &c., by George Grey, Esq. Capt. 83rd Regiment.

as occurring in places, as far south as Arrowsmith River. He likewise mentions the red sandstone about Gantheaume Bay as lying horizontally under the white sandy limestone. All these observations go to strengthen my idea of this red sandstone being the lower part of the tertiary formation.

VI. NORTH-WEST COAST.

Of the coast from Sharks Bay round the N. W. Cape to Dampier's Archipelago, I can find no published account which contains much geological information. Captain Stokes describes the coast from N. W. Cape to the rocky promontory near Regent's River as a dead flat of sand. In the Dampier Archipelago, Malus Island is said in Dr. Fitton's Appendix to Captain King's voyage, to consist of greenstone. Depuch Island is likewise said to be basaltic, or volcanic, in the same voyage, on the authority of M. Peron, p. 147. But in Captain Stokes' voyage, both Depuch Island and the neighbourhood, is said to be formed of a coarse gritty yellow sandstone, much honeycombed, p. 168. Captain Stokes also describes Barrow's Island as consisting of red sandstone. The probabilities perhaps are in favour of the rocks of the hill country of Western Australia striking out to the N. W. Cape, partly fronted and covered by the tertiary rocks.

From the neighbourhood of Dampier's Archipelago to Roebuck Bay, it appears from the accounts of

Captain Stokes* and other voyagers, that the country is an absolute flat, scarcely raised above the level of the sea, a sandy desert covered by salsolaceous plants, and fronted by lines of sand dunes running along the beach. Of such a country all we can say is, that the probabilities are in favour of its being of tertiary or still more modern formation.

VII. North Australia.

Under this appellation I shall include all the country from about Roebuck Bay to the Gulf of Carpentaria.

From Roebuck Bay to Collier Bay we have still some doubtful ground. Captain Stokes gives the following descriptions of localities in this district. "Point Emeriau, tall white cliffs springing from black rocks tinged with red," p. 83. "Cape L'Evêque, red cliffy point, 60 feet high," p. 94. "Small bight, near Point Swan, cliffs rise 70 to 90 feet, bases resting on masses of the same white sandstone as themselves, together with ferruginous rocks described by Darwin as a superficial highly ferruginous sandstone, with concretionary veins and aggregations," p. 108. "On the banks of Fitzroy river yellow sandstone cliffs, 16 feet high, and coarse red sandstone with quartz pebbles, together with a mound of loose white sand," p. 148. In King's Sound, at Port Usborne, are "rugged sandstone ranges, with dark valleys," p. 161,—"the country

* "Discoveries in Australia, &c. by Captain J. Lort Stokes, R.N."

covered with masses of sandstone," p. 168. The red and white sandstones with ferruginous concretions I should conclude to be the same as the rocks of Port Essington, which I suppose to be tertiary: but the "coarse sandstone" and the "rugged ranges" I take to belong to the rocks of the district about to be described.

This district extends from Collier Bay to Victoria River, and has been described by Captain King in the account of his first arduous and important surveying voyages round the coasts of Australia, and more recently by Captains Stokes and Grey. This district is wholly different from any other part of the north coast of Australia, which is generally low and level, whereas here we have lofty hills rising fully 1000 feet above the sea, and bold rocky headlands, with deep bays bounded by broken and precipitous cliffs. Captain King says that in passing through the eastern part of Port George IV. into Roger Strait, he landed on an island which was found to be granite, p. 189, and speaks of some islets of grey slate in Doubtful Bay. Captain Grey likewise speaks of porphyry and basalt in Port George IV., and after crossing the sandstone ranges, he describes a volcanic cone with lava and other hills of basalt in the valley of his Glenelg River, pp. 162 and 168. With these exceptions, the whole of this district appears to be composed of sandstone in thick and massive beds, apparently identical in lithological character with the palæozoic sandstones of New South Wales and Tasmania. Captain King, on coming from the eastward speaks of the character of the country about Cambridge Gulf being altogether different from that which he had hitherto seen, "instead of the low shores which prevailed from Cape Wessel, or for 600 miles, there are now irregular ranges of detached hills of a sandstone formation" (vol. i. p. 291). In La Crosse Island, he speaks of fine grained sandstone in thick beds dipping gently to the S.E., p. 282. In the Eclipse Islands (near Cape Londonderry), quartzose sandstone in large water-worn masses on basis of same rock, p. 312. Port Warrender same sandstone crusted with quartz, p. 223. Water Island, one of the Montalivet Islands, horizontal strata of hard quartzose sandstone, covered with rude blocks of the same, p. 396. The islands off Cape Bond, a heap of sandstone rocks, p. 401. Hunter River, same great sandstone beds, p. 403. And at Roe River, where the cliffs are three hundred feet high, p. 408. Also about Careening Bay (in Brunswick Bay) where the beds dip at 5° to the S.E., p. 422. Prince Regent's Inlet and river and all the neighbourhood is composed of the same sandstone in beds from six to twelve feet thick, lying horizontally, the surrounding hills being upwards of 400 feet in height, (vol. ii. p. 46). In Hanover Bay, sandstone rocks, and the hills 250 feet high (vol. ii. p. 58). Captain Grey describes the great masses of sandstone in the country around Hanover Bay, where it forms a succession of terraces and table-lands, one of which, he says, is curiously worn by degradation into a number of columns of sandstone rock, p. 97, vol. i. He mentions the same ranges about the Glenelg (p. 162), and at the farthest point he reached, he describes Stephens' Range as of sandstone, running N.N.E. and S.S.W. with lateral spurs or branches, (p. 265.)

Captain Stokes says that the height of Mount Trafalgar and Waterloo, near Prince Regent's Inlet, is about 900 feet. He describes hills of similar sandstone as running up the estuary of his Victoria River, where in Entrance Island, he says, it has a marked dip to the S.E. and at another point he speaks of hills composed of white compact sandstone, dipping S.E. at 30°. Inside Point Pearce, in this inlet of his Victoria River, is one point called " Fossil Head," from which fossils were procured, but which were unfortunately either lost or destroyed.* In a letter to myself, Captain Stokes described them as " casts of shells, not of a recent appearance." I am aware of the very slender foundation for any authoritative opinion on the age of these rocks from the data here given, but the

---

* I examined the boxes of rocks brought home by Captain Stokes, which were lying at the Admiralty, but did not succeed in finding any fossils. Mr. Darwin had previously examined them, but had been equally unsuccessful. The lumps of rock from the neighbourhood were like the palæozoic sandstone of New South Wales.

balance of probabilities appears to me to be in favour of their belonging to the same palæozoic formation which is found so largely in New South Wales, and of which we have a trace in Western Australia. I have, however, left them doubtful in the accompanying map.

We have already seen Captain King's opinion of the uniformity of character of the remainder of the north coast from the indentation of Cambridge Gulf to the Gulf of Carpentaria, the land being generally low and the rocks loose sands with ferruginous concretions. We shall, however, see reason to believe that in the interior of the country, the massive sandstones of supposed palæozoic age, stretch across in a lofty range from the neighbourhood of Stokes' Victoria River to that of the Gulf of Carpentaria. I will first describe the coast formation. The only point where I saw this myself, was at Port Essington. Here it was a red and white ochreous sandstone, sometimes rather argillaceous when it was firm and compact, but generally soft and friable. The red and white colours were sometimes confined to different portions of the cliff, but at others, they were intermingled in blotches. Wherever the stratification was discernible, it was always horizontal, but sometimes a cliff of twenty or thirty feet high would consist of light powdery sandstone, without any marks of bedding. It frequently contained ferruginous concretions, and these sometimes occurred in such abundance as to occupy the entire thickness exposed, and many of the points and headlands consisted of a pile of these large concretions of iron ore looking like a heap of slags from an iron furnace. These concretions were black inside, but frequently red outside; when struck with a hammer they emitted a metallic sound, when broken open they seemed to be made up of small grains of quartz, with others of some mineral with a glittering metallic surface. They did not affect the magnet, but when tested by prussiate of potash and extract of galls, they appeared principally an ore of iron, while under the blowpipe it coloured a bead of borax a purple black, from which I conclude it is a compound of iron and manganese. I searched in vain for any organic remains in any of the rocks; but on visiting Port Phillip I was struck with the exact lithological resemblance in all these characters, which the tertiary rocks on the east side of that port bore to those of Port Essington. From this circumstance, from their perfect horizontality, from their little height above the sea, and from the incoherent and slightly consolidated state of these rocks of the north coast, I was induced to look upon them as likewise of tertiary age. I have, however, left the point doubtful in the accompanying map.

Captain King describes the cliffs about Anson Bay and Cape Ford and thence to Cape Dombey and Port Keats as of moderate height, level, made of dark red sandstone, or tenacious clay with ironstone pebbles, &c. (p. 274 to 277). The coast south of Melville's Island as low deep red coloured cliffs all along the land (p. 147). Lethbridge Bay in Melville's Island argillaceous sandstone, upper part red, lower white (p. 108). East of Port Essington, he describes the coast all about Goulbourn's Island as having cliffs forty feet high, upper stratum red indurated clay, lower whitish pipeclay (p. 64). He speaks of an argillaceous cliff, bright yellow, at Copeland Island (p. 77). And says generally that the coast was exactly of the same character thence all the way to Cape Wessel.

From Dr. Fitton's Appendix to Captain King's Voyage, we gather that there is granite in the country about Melville Bay and Cape Arnhem, that at Morgan's Island and an islet in Blue Mud Bay there is chinkstone and trap, and that in Caledon Bay there is likewise granite, but that these rocks are covered with a brown hæmatite and concretionary ironstone, and that all along Limmen's Bight there is a range of low hills of which the base is " primitive ;" the upper part reddish sandstone. Groote Eylandt is said to be composed principally of sandstone and conglomerate. I believe that here as on the south coast on each side of the Great Australian Bight, we have tertiary rocks resting on granite, and the other rocks formerly called primitive.

Of the interior of this tract of country we know nothing except from the accounts of the adventurous

Leichhardt. At the southern part of Limmen's Bight, just south of Wickham's River, he speaks of a range of hills of "white sandstone, baked, resembling quartzite, dipping at small angles to the south, and striking east and west," (p. 426 to 434). He meets with high sandstone ranges steep to S.E., and sloping gently to N.W., rock generally white, between Wickham and Roper's Rivers, (p. 437). He then describes a continually ascending country of sandstone as he goes up Roper's River, with indurated clay and distinct stratification till near latitude 13°, and longitude 132° and 133° he arrives at a high table land of massive sandstone horizontally stratified, which he says at one place seemed to be "literally hashed, leaving the remaining blocks in figures of every shape," (p. 475). This country seems to have been the counterpart of that traversed by Grey in penetrating to his Glenelg River, (see ante p. 70). In longitude 134°, latitude 14° 20′ he speaks of basalt surrounded by high hills of sandstone (p. 458), and in latitude 13° 50′, longitude 133° 33′ he says the sandstone dips S.W. (p. 469). Just previously (p. 468) he speaks of a baked sandstone, in some pebbles of which he found "impressions of bivalves, ribbed like a cockle." These may have been spiriferæ or productæ,— which I have often heard spoken of as " cockles," in New South Wales and Tasmania.

From the high table land, 1800 feet, to which he attained south of Port Essington, he looked down on the flat country around the Alligator Rivers, to which he found some difficulty in descending, from the precipitous nature of the ground. Below the broken sandstone cliffs he found a coarse granite passing into sienite, and a high range of pegmatite descended from the table land far into the valley from east to west, (p. 487). In the flat country from the foot of these lofty hills to Port Essington he meets with nothing but "ironstone and conglomerate with pebbles and pieces of quartz" (p. 490), "rocky sandstone hills with horizontal stratification" (p. 507), "hills one or two hundred feet high, with clayey ironstone" (p. 529), evidently the same rocks as those stretching along the coast. The massive sandstones forming the upper part of the high table land, are almost certainly the same formation as that forming the country between Victoria River and Collier Bay.

We have now merely to describe the head and eastern side of the Gulf of Carpentaria. Leichhardt travelled round the head of the Gulf without meeting with any fresh water river of importance, though he crossed several estuaries of some size which were full of sea water.* He speaks of a range of hills of

* On heading round the salt water they never found any thing but the most insignificant stream of fresh water running into it, although the bed of the river was often large and wide, evidently to allow of the passage of large occasional floods. All the rivers of the north coast, as appears from Stokes's accounts of his Victoria, Adelaide, Albert River, &c. have the same character.

sandstone, horizontal strata deeply fissured in long.
136°, lat. 16° 10′; but on either side of them he
only mentions red ironstone (p. 415). Near the
River Van Alphen we read of white sandstone hard,
flaggy, and horizontally stratified (p. 396). Thence
to Van Diemen's River psammite (grains of quartz
in a clayey base) is said to form the basis of the
country with ironstone and iron sandstone occurring
occasionally (pp. 351, 375, 379, 387).

Captain Stokes in speaking of the Wellesley
Islands describes Bountiful Island as made of sand
and ironstone, ferruginous sandstone, and reddish
sandstone, very much honeycombed (p. 267);
Bentinck Islands, ironstone cliffs (p. 277), Flinders'
River red ferruginous rock (p. 299).

In Dr. Fitton's Appendix to Captain King's
Voyage there is described a calcareous sandstone of
recent concretionary formation as found on Sweers
Island, one of the Wellesley group. The whole
eastern coast of the Gulf of Carpentaria has been
always described by Flinders, King, and Stokes as
one great flat of sand, scarcely raised above the level
of the sea, of which flat I saw the extremity in the
south side of Endeavour Straits. The sand there
was white and appeared to be loose. The only
rock mentioned is a calcareous sandstone of recent
concretionary formation, said in Dr. Fitton's Appendix to King to have been brought from the mouth
of Coen River, which is a little south of the Batavia
River, marked in the accompanying map. That

this great plain extends far into the interior round the head of the Gulf is apparent, from no hill having been seen rising beyond it either by Leichhardt or by Stokes when he reached what he calls the " Plains of Promise."

GENERAL CONCLUSIONS.

I would now call the reader's attention to some general conclusions from the foregoing details. We may first notice the simplicity and uniformity of the geology of Australia, when looked at on the great scale. The strike of the rocks and the direction of the principal chains of hills, is with one exception the same throughout the country, namely, north and south. This is the case with the great chain of the Eastern coast, with its subordinate ranges that traverse the interior of New South Wales and Australia Felix, such as the Hervey Range, Balloon or Taylor Range, and Peel Range, on the Lachlan river; Mount Byng Range, the Pyrennees and the Grampians in Australia Felix. The chain of South Australia runs nearly due north and south, as does the ranges visited in the interior by Sturt,—Stanley Range and Grey Range. The ranges of the hill country of Western Australia, have the same direction from King George's Sound, to the neighbourhood of Sharks Bay. In North Australia, however, we find a great change of strike, there are several dips to the S.S.E. and S.W. mentioned, and the direction of the hills is frequently spoken of as east and west, which seems indeed to be the general bearing of the hilly land from Collier Bay to the Gulf of Carpentaria. All the large spaces between and around these ranges of old rocks are occupied by great plains, formed of tertiary rocks in a horizontal position, which, as we have seen, spread unbroken and unchanged sometimes for hundreds of miles.

The total absence of any rocks of an age intermediate between the palæozoic and the tertiary, so far as is at present known or appears probable, is another general result, worthy of remark, both as producing simplicity of structure, and for other reasons.

We are naturally led from these two results to two speculations, one as to the physical geography of the great tract still unexplored in the centre of Australia,— another as to the past geological history of the country.

Let us first consider what is probably the character of the unexplored interior of Australia. When we look at the great tertiary plain into which the Murray and other rivers of New South Wales fall on their road to the sea, at the immense tertiary plateau round the Great Australian Bight, that spreads horizontally for an unknown distance into the interior, and which appears to sweep up to the very borders of the hill country of Western Australia (see ante p. 63, and 67, the account of the sand plains), at the great flats, probably tertiary, that extend from the N.W. coast between lat. 17° and 22°, likewise to an unknown extent into the interior, and to the plains round the Gulf of Carpentaria, which seem equally to spread out towards the same centre, and combine these facts with the existence of an immense desert plain in that central interior as visited and described by Captain Sturt, we are almost irresistibly induced to look upon all these low and level tracts as but separate portions of one immense plain, occupying by far the larger portion of Australia, like a great sea of low and level land, with the hilly districts rising from it like islands.

The known facts as to the climate and meteorology of Australia would confirm this supposition to a very considerable extent. The climate of all the colonies and all the coasts of Australia is remarkable for its dryness,—for its long droughts, and consequently for the little permanent water that is to be found in the country either as lake, pond, or river. In New South Wales, several periods of one, two, or three years have already occurred since its colonization in which not a drop of rain has fallen. Great rains indeed fall occasionally, apparently in all parts, but they are irregular and partial, form sudden floods of great magnitude and soon rush off the land into the sea, without leaving any large store of water either on the surface or in the depths of the earth. The only large rivers that are at present known are those which flow off the western slope of the Great Eastern Chain in New South Wales, and of these the only one having a permanent stream of water is the one which rises from the only mountain summits that are permanently covered with snow,—namely, the river Murray from the Australian Alps. The whole remaining coast of Australia, has now been carefully surveyed without the mouth of any large or important stream having been discovered.

In all the Australian colonies they have during the summer season occasional blasts of what is called the "hot wind." This is often a strong and steady breeze before which the trees bend and which blows up clouds of dust and sand, and sometimes of gritty particles large enough to strike with painful acuteness on the face. Instead of a cooling breeze, however, the breath of this wind is like the blast from a fiery furnace. The leaves of the trees seem to wither and shrink up under its scorching influence, things exposed to it, such as the handles of doors, rocks, &c. become so hot as to be almost painful to the grasp of the hand. The perspiration instantly dries upon the face, the lips parch, and the skin tingles. The sky though clear of clouds assumes a hazy aspect through which the sun glows like a copper ball. Doors and windows are instantly closed, and men and animals all seem to seek shelter from the painful influence of this scorching blast.*

* Though exceedingly disagreeable there is however nothing debilitating or immediately injurious in these "hot winds," as, if a person is obliged or resolved to endure them, I can say from personal experience that it is possible to walk and to work the whole day long in the very teeth of them.

This hot wind always blows from the interior of the country. In New South Wales and Tasmania it blows from the N.W. and rarely lasts more than two days. It is generally succeeded by a sudden squall from the southward, which brings up clouds and often rain. During the hot wind the thermometer rises to 100° or even 115° in the shade, but with the southerly squall there is sometimes a sudden fall of full 40° in the course of a half or even a quarter of an hour. Near Port Phillip the hot wind comes from the north or even the N.E. the hot stream being probably deflected by the high lands of the Eastern Chain. In South Australia the hot winds come from the north, and sometimes continue as long as nine days. Many of the houses in Adelaide have double walls with a space between on their northern sides to ward off some of its influence. The longer duration and I believe the greater intensity of the hot wind here is evidently due to the close proximity of the central desert and that the mountain chains, such as they are, run parallel to the course of the wind and thus do little if anything towards sheltering the colony from its influence. In Western Australia the hot wind, or as it is there called the "land wind," blows from the N.E., but from my own experience, though sufficiently parching, they have not the fierceness and intensity which they have in Southern and Eastern Australia. Captain King describes somewhat similar effects on the N.W. coast, as arising from a wind blowing from the interior of the country or the S.E., while at Port Essington every wind that does not blow over the neighbouring sea is hot and overpowering. These hot winds, especially those felt in the south-eastern part of Australia, can have no other origin, than a current of air blowing over some great expanse of burning desert. Captain Sturt in his last expedition, seems to have penetrated into the very "nest and hot-bed" of these hot winds. In addition to his published accounts of this burning sea of sand ridges and ironstone flats to which he penetrated, I recollect his telling me when I met him in South Australia in 1845, on his return from his arduous journey, that the sands were so hot that if a lucifer match were let fall on them it instantly took fire, and that the blasts of wind were sometimes of such an intensity of temperature that they could not face them, but were obliged to stoop the head till they passed by. The existence of these hot winds is incompatible with the idea of any expanse of water or of any mountain chain of importance in the interior of Australia, while it exactly agrees with our conclusions drawn from geological evidence, of that interior being for the most part a low and arid plain.

The total absence of large rivers is another important fact in accordance with this supposition. The northern half of Australia is within the influence of the trade wind that blows throughout the year from the E.S.E. In the winter of that hemisphere

(May to September) the trade winds are confined to the Tropics, and blow all along the north-eastern and the northern coast of Australia. At that season calms, and light, and variable airs prevail all along the north-west coast for a distance of at least 200 or 300 miles from the shore. The trade wind which blows "home" on the eastern coast is lifted by the ascending current that rises from the heated plains of the interior, and does not descend to the sea again on the north-west till after it has passed beyond them for that distance. In the summer (November to March) the sun draws the trade wind as far south as 30° or 35°, so that easterly winds prevail all along New South Wales, and frequently blow through Bass's Straits and along the southern coast of Australia. At that time the sun being perpendicular over Northern Australia heats it to such an extent that instead of mere calms on the north-west coast the wind rushes from the north-west to the land, the current is reversed instead of suspended, and occasional winds and gales from the west and north-west prevail all along the northern coast. That season is thence called that of the north-west monsoon.* These winds bring heavy thunder

* This north-west monsoon being, I believe, first originated by the heated surface of Australia, is strongest and steadiest in the Northern Indian Ocean, south of Java and the other islands up to New Guinea. Having once begun to blow, however, the other islands, New Guinea, New Hebrides, the Solomon Islands, New Caledonia, &c. have sufficient heated surface to draw on and

showers, and during their prevalence, is the only time throughout the year when rain was known to fall at Port Essington. In the winter the trade wind that blows on the north-east coast, of course, brings much moisture with it. The summits of the eastern chain catch and precipitate much of this moisture, their peaks being frequently clothed with clouds, and a multitude of small rills and brooks run down their eastern slopes into the sea. After passing over this ridge, however, and being thus partially drained of its moisture it would appear that the south-east trade wind meets with no other range, at least of equal or greater altitude, or we should have streams like these combining to form a river that must come out upon the coast. Instead of this its behaviour exactly accords with its passing over plains and burning deserts, where it is heated and rarified instead of cooled and condensed, and thus becomes capable of taking up more moisture instead of being obliged to part with that which it possessed.

In the same manner during the summer season, if any great mountain range existed in the interior of Australia, it must attract and condense the moisture with which the north-west wind is loaded, precipitate it into rain, and send it back to the north-west or to some other coast in the shape of a

and continue this north-west wind up to the Feejee Islands, but it blows there in a more partial and interrupted manner than in the Indian ocean.

river of some magnitude, greater than any of the inconsiderable streams which alone have been discovered. It appears to me that the only chance of any exception to this flat and barren state of the interior being discovered is that springing from the possibility of the ranges of palæozoic formation between Collier Bay and the Gulf of Carpentaria, increasing in magnitude in the interior or loftier ranges springing up parallel to their strike, and that from these springs and streams may arise sufficient to fertilize the adjacent country, although not enough to form a river large and full enough to force its way to the coast.

It may, perhaps, illustrate my meaning in the preceding remarks on the climate of Australia if we briefly compare it with the part of South America, in about the same latitudes. There we have a continent exposed to the perennial E.S.E. wind from 10° and 15° of latitude, to 25° and 35°, and to westerly winds almost equally constant south of 35°. In South America, however, the principal mountain chain, the Andes, is on the western coast instead of the eastern as in Australia. The consequence is that south of lat. 35°, or in the region of the westerly winds we have a desert on the eastern side of the range in the plains of Patagonia, watered only by streams that rise in the perpetual snow of the Andes, while on the western side the hills and the belt of land between the mountains and the sea with the adjacent islands from Cape Horn to Valparaiso are clothed with the densest forest, and often, as in Chiloe, drowned in perpetual torrents of rain. North of 30° or in the region of the easterly trade winds the exact opposite is the case. Here the whole country of Brazil and the neighbouring territories east of the great chain are clothed with magnificent forests, refreshed by frequent rains, and traversed by magnificent rivers that, springing from the snows of the Andes, are kept up and replenished, and continually added to by streams flowing from every portion of the country, while, on the western slope of the great chain in the narrow belt of Peru occur great deserts without a drop of water, and rain is seldom, or in many places, never known to fall.* In each case the wind loaded with the moisture it has taken up in passing over the ocean is effectually drained of it in crossing the cold altitudes of the mountain chain. It is obvious that if we changed the position of the Andes while the winds remained the same, if the mountain chain ran along the eastern coast of South America, the position of the fertile and desert tracts would be reversed. Patagonia would be covered with forest, while Brazil would be a hot, desert, and arid region, with a mere strip of watered and fertile country along the eastern coast. Now, this supposed case, is exactly that of Australia, with the difference, that as the eastern chain of Australia

* Among other travels in South America which describe these facts, the reader's attention is particularly directed to Mr. Darwin's " Journal of a Naturalist."

is so much lower and less important than the Andes, its effect is not so decided and important; what effect it does exert, however, is precisely of the same kind and in the same direction.

Our second speculation, is on the past geological history of Australia. Many geologists have been struck with the entire absence of all the " secondary" formations in Australia, and with the analogies between the fossil flora and fauna of our European oolitic series, and those now found living in Australia and the Australian seas. We have seen in the preceding pages that we have in Australia a group of very old rocks, now more or less completely metamorphosed, and associated more or less intimately with granitic rocks; on these we have a palæozoic formation very largely developed, but we have at present no knowledge of the exact relation between the palæozoic and the metamorphic series. Above the palæozoic series there is an absolute gap, a total deficiency of all other stratified rocks whatsoever, so far as is at present known, except those belonging to a tertiary formation, which from the very recent aspect of its fossils and their resemblance to existing forms, I believe to be a very modern one. Now as there seems to be no rocks of the age of our great oolitic and cretaceous formations, they must either have been altogether destroyed and denuded which is unlikely, or they can never have been deposited. This circumstance inclines us to believe, that the country may, most probably, have existed as dry land during the oolitic and cretaceous periods, if so, is it not possible that its present fauna and flora may be in some way the descendants and representatives of the fauna and flora that in the oolitic period was common to the whole earth? We have in our oolitic rocks in Europe fossil trigoniæ and other shells, fossil plants allied to zamiæ and cycas, and a fossil marsupial animal. As is well known to every one, Australia is the head quarters of the marsupial animals of the present day, zamiæ and cycadeæ abound in its forests, growing in the burnt up sandy plains, covered with ironstone pebbles under lofty slender leaved eucalypti, and the Australian coasts are the only localities where trigoniæ are found now living in the world. I believe the parallelism of the living flora and fauna of Australia, and that fossil in the oolites might be carried still farther, if my knowledge of each were greater and more exact: but in what has been said above, we find that there is at least a generic or family resemblance in some of the land animals, some of the land plants, and some of the animals of the sea.

We have therefore two reasons; namely, the absence of marine formation of the oolitic age, and the possible descent of some of the animals and plants from those that lived at that period; for supposing that after the deposition of the palæozoic rocks, what is now Australia was raised into dry land, and that some portion or portions of it at all

events have ever since remained above the level of
the sea. During some tertiary period, however, it
is clear that it was again partially submerged, that
a sea of no great depth extended over by far the
greater portion of Australia, in which sea was
deposited the beds of limestone and ferruginous
sandstone, and the loose and incoherent sands
that form the great plains of which *we know* so
large and *I believe* so much larger a part of the
country is composed. During this tertiary period
we should have one long island, or chain of islands,
running along what is now the great eastern chain
of Australia; a similar island, or islands, but of
much less extent, formed of the chain of South
Australia, with many smaller groups between the
two in what is now Australia Felix. These latter
were probably added to at this period by a number
of volcanic cones, that probably arose above the
level of the sea from the volcanic foci that certainly
found vent in its bed. There would probably be
another group of islands formed of the high lands
of Western Australia, and again another running
probably east and west where we now have high
land on the north-west coast, and in the interior,
south of Port Essington. After this partial sub-
mergence of the country, it has again been elevated
above the sea, and left in its present condition.
This geological history would account for the
present physical features of the country, and likewise for the specific differences which may be

detected under the strict generic resemblance of the fauna and flora of different parts of Australia.

The mountain chains of old rocks have not only passed through the destructive plains of the sea level, and been everywhere subject to the wearing action of its breakers, which, in addition to the dislocating and disturbing forces acting on the interior, must have destroyed and removed so large a portion of their mass; but they must for long, long ages have stood the wear and tear of the wind and the rain, and the other atmospheric causes of degradation and disintegration which would sufficiently account for their remarkably furrowed character, their deep erosion into valleys and ravines, which even where the rocks are but little disturbed, is so striking a characteristic of many of the old rocks of this country.* On the contrary, the tertiary rocks having been merely elevated en masse, and that, geologically speaking, at so recent a date, form level or gently undulating and unbroken plains, which are an equal characteristic of the Australian land.

I believe, partly from my own observation, partly from the oral information or the published accounts of others, that the plants and animals of Australia, though everywhere similar, or to speak strictly, though every where of the same order and family, and commonly of the same genus, yet vary in species and sometimes in genus, not simply with

* See pp. 23 and 24.

the latitude and the height above the sea, but also with the distance of the locality in longitude. There are, for instance, very few mammals, I believe, and not many birds even, that are common to Western Australia and New South Wales, although the climate generally and the rocks and every other circumstance of the two countries is frequently the same. The eucalypti of the one country, which give the general feature to the forests and the scenery, or which are the most abundant, are not the same eucalypti which are most abundant in the other. There are kangaroos in both, but not the same species of kangaroos; the difference in the common parroquets and other birds that one sees continually about one, is most striking whenever one passes from one colony to another. This difference is most striking in passing from Sydney to Swan River, but it is quite perceptible in moving from any colony to another, Port Phillip to Adelaide, or Van Diemen's Land, for instance. The same is the case even with the marine mollusca to a very considerable extent, more especially perhaps the littoral species. I believe I should be correct in saying that there is a greater difference found in the species of the commoner plants and animals (including birds) in passing from the neighbourhood of Sydney to that of Adelaide, and certainly in going from Sydney to Swan River, though the latitude and the climate is so nearly the same, than in going from Sydney along the N.E.

coast, nearly up to Torres' Straits, though we then pass from the latitude of 34° nearly to the neighbourhood of the equator.*

Now our hypothetical geological history would account for this. We may suppose that previous to the depression that let in the tertiary sea over Australia, that country, (and perhaps a large portion of the adjacent surface of the earth), was covered by trees, plants, and animals, whether terrestrial or freshwater, and its coasts bordered by marine animals, all of a common type, that their species and genera varied only with the variation in latitude, depth, altitude, nature of the soil, &c., or all those circumstances that go to make up the "climate" of a place. When this country, however, was broken up into islands, it is probable that each island or group of islands would preserve some species or genera of animals or plants that would be lost in the others. Lapse of time would tend still farther to modify these different assemblages by the extinction of some of the old species, and the introduction of new ones, until at last the more distant groups exhibited a specific contrast while they had a generic resemblance to each other. We should thus have a reason for a certain peculiarity of the fauna and flora of each tract of country;—for the eastern coast which was once one long island or group of islands, having plants and animals of the

* About the same change of latitude as from the south of Spain to Sierra Leone.

same species throughout its extent, which did not occur on the western, and *vice versâ*. Each group of islands when raised so as to be all connected by dry land, would be prepared to act the part of a specific, or even perhaps a generic centre, and its inhabitants would descend on the newly acquired plains, to reconquer and reclaim their old territories, or to occupy the new. Some of the islands, however, having remained as islands, might preserve their peculiarities, which would account, in the instance of Tasmania, for its peculiar possession of those extraordinary animals, the Dasyurus ursinus and Thylacinus cynocephalus, (the " devil" and the " tiger" of the colonists). Others might still be separated by impassable deserts from the rest, as in the instance of Western Australia, where the peculiarities of the fauna and flora are more marked than in the other colonies.

THE END.

G. NORMAN, PRINTER, MAIDEN LANE, COVENT GARDEN.

Printed in the United States
By Bookmasters